Impact of Climate Change on Mental Health and Well-Being

Debabrata Samanta
Rochester Institute of Technology, Kosovo

Muskan Garg
Mayo Clinic, USA

A volume in the Advances in Psychology, Mental Health, and Behavioral Studies (APMHBS) Book Series

IGI Global
PUBLISHER of TIMELY KNOWLEDGE

Published in the United States of America by
IGI Global
Information Science Reference (an imprint of IGI Global)
701 E. Chocolate Avenue
Hershey PA, USA 17033
Tel: 717-533-8845
Fax: 717-533-8661
E-mail: cust@igi-global.com
Web site: http://www.igi-global.com

Library of Congress Cataloging-in-Publication Data

Names: Samanta, Debabrata, 1987- editor. | Garg, Muskan, editor.
Title: Impact of climate change on mental health and well-being / edited by Debabrata Samanta, Muskan Garg.
Description: Hershey, PA : Information Science Reference, 2024. | Includes bibliographical references and index. | Summary: "Climate change has far-reaching consequences beyond its environmental impact. It also significantly affects social and mental well-being, both at individual and community levels. Addressing the social and mental well-being impacts of climate change requires a multi-faceted approach that includes both mitigation and adaptation strategies"-- Provided by publisher.
Identifiers: LCCN 2023055022 (print) | LCCN 2023055023 (ebook) | ISBN 9798369321775 (hardcover) | ISBN 9798369321782 (ebook)
Subjects: LCSH: Climatic changes--Psychological aspects. | Mental health--Environmental aspects. | Well-being--Environmental aspects.
Classification: LCC BF353.5.C55 I47 2024 (print) | LCC BF353.5.C55 (ebook) | DDC 155.9/15--dc23/eng/20231218
LC record available at https://lccn.loc.gov/2023055022
LC ebook record available at https://lccn.loc.gov/2023055023

This book is published in the IGI Global book series Advances in Psychology, Mental Health, and Behavioral Studies (APMHBS) (ISSN: 2475-6660; eISSN: 2475-6679)

British Cataloguing in Publication Data
A Cataloguing in Publication record for this book is available from the British Library.

For electronic access to this publication, please contact: eresources@igi-global.com.

Advances in Psychology, Mental Health, and Behavioral Studies (APMHBS) Book Series

ISSN:2475-6660
EISSN:2475-6679

Editor-in-Chief: Harish C. Chandan, Argosy University, USA

MISSION

The complexity of the human mind has puzzled researchers and physicians for centuries. While widely studied, the brain still remains largely misunderstood.

The **Advances in Psychology, Mental Health, and Behavioral Studies (APMHBS)** book series presents comprehensive research publications focusing on topics relating to the human psyche, cognition, psychiatric care, mental and developmental disorders, as well as human behavior and interaction. Featuring diverse and innovative research, publications within APMHBS are ideally designed for use by mental health professionals, academicians, researchers, and upper-level students.

COVERAGE

- Anxiety
- Neurological Disorders
- Treatment & Care
- Human Interaction
- Cognition
- Mental Health & Disorders
- Substance Abuse
- Psychiatry
- Socialization
- Developmental Disorders

IGI Global is currently accepting manuscripts for publication within this series. To submit a proposal for a volume in this series, please contact our Acquisition Editors at Acquisitions@igi-global.com or visit: http://www.igi-global.com/publish/.

The Advances in Psychology, Mental Health, and Behavioral Studies (APMHBS) Book Series (ISSN 2475-6660) is published by IGI Global, 701 E. Chocolate Avenue, Hershey, PA 17033-1240, USA, www.igi-global.com. This series is composed of titles available for purchase individually; each title is edited to be contextually exclusive from any other title within the series. For pricing and ordering information please visit http://www.igi-global.com/book-series/advances-psychology-mental-health-behavioral/110200. Postmaster: Send all address changes to above address.

Titles in this Series

For a list of additional titles in this series, please visit:
http://www.igi-global.com/book-series/advances-psychology-mental-health-behavioral/110200

African Womanhood and the Feminist Agenda
Maxwell Constantine Chando Musingafi (National University of Lesotho, Lesotho) and Chipo Hungwe (Midlands State University, Zimbabwe)
Information Science Reference • © 2024 • 280pp • H/C (ISBN: 9798369319994) • US $230.00

Promoting Mindfulness and Well-Being with Indian Psychology
Anuradha Sathiyaseelan (Christ University, India) and Sathiyaseelan Balasundaram (Christ University, India)
Information Science Reference • © 2024 • 300pp • H/C (ISBN: 9798369326510) • US $245.00

Challenges and Opportunities for Women, Parenting, and Child Development
Nava R. Silton (Marymount Manhattan College, USA)
Information Science Reference • © 2024 • 300pp • H/C (ISBN: 9798369312858) • US $230.00

Global Applications of Indian Psychology Therapeutic and Strategic Models
Anuradha Sathiyaseelan (Christ University, India) and Sathiyaseelan Balasundaram (Christ University, India)
Information Science Reference • © 2024 • 325pp • H/C (ISBN: 9781668497784) • US $265.00

Combining Exercise and Psychotherapy to Treat Mental Health
Lara Carneiro (United Arab Emirates University, UAE) and Felipe Schuch (Universidade Federal de Santa Maria, Brazil)
Medical Information Science Reference • © 2024 • 359pp • H/C (ISBN: 9781668460405) • US $325.00

Childhood Developmental Language Disorders Role of Inclusion, Families, and Professionals
Dimitra Katsarou (University of the Aegean, Greece)
Information Science Reference • © 2024 • 424pp • H/C (ISBN: 9798369319826) • US $280.00

IGI Global
PUBLISHER of TIMELY KNOWLEDGE

701 East Chocolate Avenue, Hershey, PA 17033, USA
Tel: 717-533-8845 x100 • Fax: 717-533-8661
E-Mail: cust@igi-global.com • www.igi-global.com

To my parents Mr. Dulal Chandra Samanta, Mrs. Ambujini Samanta, my elder sister Mrs. Tanusree Samanta, Brother-In-law Mr. Soumendra Jana and daughter Ms. Aditri Samanta.

-Debabrata Samanta

To my loving family, whose unwavering support and endless encouragement have been my greatest source of strength. You are my inspiration, my motivation, and my reason for never giving up. This book is dedicated to you, with boundless gratitude and unending love.

-Muskan Garg

Editorial Advisory Board

M. S. Mekala, *Yeungnam University, South Korea*
Regonda Nagaraju, *Sreyas Institute of Engineering and Technology, Hyderabad, India*
Neelamadhab Padhy, *GIET University, Gunupur, India*
Indrajit Pan, *RCC Institute of Information Technology, India*
Yohan Park, *Keimyung University, South Korea*
Binod Kumar Pattanayak, *Soa University, India*
Suplab Podder, *Dayananda Sagar Institutions, Bangalore, India*
Gurunath R., *Dayananda Sagar College of Arts, Science, and Commerce, Bangalore, India*
Gurunath Ramachandra, *Christ University, India*
Poonam Rani, *Netaji Subhash University of Technology, New Delhi, India*
Ahmed Nabih Zaki Rashed, *Menoufia University, Egypt*
Joel J. P. C. Rodrigues, *Federal University of Piauí, Brazil*
Rita Roy, *GITAM Institute of Technology, India*
I. G. Samusev, *Amur State University, Russia*
A. Y. Seteikin, *Amur State University, Russia*
M. Shivamurthaiah, *Garden City University, India*
Prakash Kumar Singh, *Rec Mainpuri, India*
P. Sivakumar, *PSG College of Technology, Coimbatore, India*
M. Sivaram, *Lebanese French University, Erbil, Iraq*
Ming-Hour Yang, *National Central University, Taiwan*
P. Yupapin, *Ton Duc Thang University, Ho Chi Minh City, Vietnam*

Table of Contents

Detailed Table of Contents

There can be no health or sustainable development without mental health underscores the inseparable link between the SDGs and mental health. Consequently, achieving the SDGs is a pathway to achieving mental health. Tourism stands as a significant contributor to positively impacting the mental health and well-being of communities. Bamboo, known as 'Green Gold' due to its carbon sequestration capabilities and other eco-friendly traits, aids in reducing carbon emissions and achieving the SDGs. The researcher has formulated a model aimed at achieving the SDGs while addressing climate change concerns and promoting well-being.

Eco-anxiety refers to chronic fear of environmental doom due to climate change and ecological harm. It stems from lived experiences, media coverage, scientific reports, and government inaction. Climate anxiety is rising, especially among youth. Exposure to climate disasters and species/habitat loss can lead to climate grief. Eco-anxiety can manifest as fatalism, avoidance, activism, or compulsions. It reflects conflict between protecting nature and feelings of helplessness. Chronic climate distress heightens risks of depression, PTSD, panic, and substance abuse. Addressing eco-anxiety requires climate education and communication, mental health resources, government climate action, and research on coping strategies. Recognition of eco-anxiety establishes climate mitigation as imperative for collective psychological wellbeing.

This chapter delves into the complex ethical dimensions surrounding the intersection of climate change and mental well-being. As global environmental shifts increasingly impact the psychological health of individuals and communities, ethical considerations become paramount in guiding effective and culturally sensitive interventions. The chapter navigates the complexities of integrating Indigenous perspectives, fostering global collaboration, and balancing immediate mental health needs with long-term sustainability goals. Case studies and ethical dilemmas offer real-world insights into the challenges faced by practitioners, highlighting the need for nuanced decision-making processes. This chapter aims to enhance the resilience, fairness, and ethical foundation of addressing the significant effects of climate change on mental health and well-being by analyzing and tackling these ethical considerations.

For the majority of people, moving abroad is already a difficult decision that brings on worry. Unfortunately, the consequences of extreme climate change are forcing people to migrate throughout the globe, causing suffering. Consequences include soil erosion, natural disasters, sea level rise, and agricultural damage. An unstoppable chain reaction is sparked by these impacts. It starts by causing environmental issues, which grow into biological issues, which directly and indirectly impact people's bodies, minds, and social structures. Furthermore, if nothing is done to stop it, this kind of displacement isn't decreasing at all and will just continue to rise. This chapter aims to provide accurate information about migration brought on by climate change, with a focus on the psychological effects; however, it also briefly discusses the physical and socioeconomic ones.

This chapter explains how the financial stability of the economy is threatened by the effects of climate change. Since research and statistics on the subject are still deemed to be shallow and inadequate, the purpose of this chapter is to address the lack of understanding regarding the extent of the effects that climate change poses to the financial industry. A wealth of information about the many economic sectors and systems impacted by climate change can be gained by investigating several economic models, such as GEM-E3-ISL and GINFORS.

Preface

Climate change is a worldwide force that deeply affects our social and emotional well-being and is no longer limited to scientific talks or environmental debates. As the editors of this extensive reference work, *Impact of Climate Change on Mental Health and Well-Being*, we set out to analyse and shed light on the significant—yet frequently disregarded—effects of climate change on society's psychological makeup.

Debabrata Samanta and Muskan Garg's partnership has been crucial in creating a story that explores the complex relationships between the significant impacts on our mental and social landscapes and the changing environment. We address the difficulties that people and communities around the world face as we examine the multi-layered complications that result from this intersection in these pages.

This book is more than a collection of data and information. In the face of environmental upheavals, it serves as a call to action and a road map for achieving resilience, adaptation, and community empowerment. It is evidence of the concerted effort needed to address the extensive effects of climate change on human welfare from a variety of societal sectors, including academics, policymaking, education, and beyond.

We have designed this reference material with a broad readership in mind, understanding that the effects of climate change on mental health cut beyond academic fields and boundaries. All readers, from scholars to practitioners, legislators to well-informed citizens, instructors to business experts, should find something of value on each page.

This book is a compilation of theoretical frameworks and empirical investigations for researchers and academics. Graduate and postgraduate students will discover a plethora of information to deepen their comprehension of this important intersection in these pages. Policymakers, government officials, and experts in the subject will find insights and tactics that may be put into practice to help guide their decisions. This is a useful tool that librarians and information professionals can use to help their clients.

Teachers and educators can use this to help students gain a deeper understanding of how climate change affects mental health. Meanwhile, readers who are generally well-informed, professionals in the field, book clubs, reading groups, and readers from around the world will find stories that connect the dots between climate science and real-world experiences.

This book is a call to participate, talk about, and act rather than a conclusion. The reader is cordially invited to peruse these pages, take in the observations, and work with us to tackle the urgent issues brought about by the mutually reinforcing relationship between climate change and our psychological and social welfare.

ORGANIZATION OF THE BOOK

In this edited reference book, *Impact of Climate Change on Mental Health and Well-Being*, we present a diverse array of chapters authored by distinguished experts, each delving into the intricate intersections between climate change and various facets of human life.

Chapter 1 focuses global economy have suffered greatly as a result of the COVID-19 epidemic, which has affected people from all walks of life, including those who live in cosmopolitan cities. Bangalore is a prime example of cosmopolitan living due to its multicultural population, which is made up of people from different ethnic backgrounds, languages, cultures, and customs. The purpose of this study is to investigate how the COVID-19 outbreak has affected Bangalore's "commoners.". Our study examines how the pandemic has affected people's quality of life in a cosmopolitan city, focusing on things like income, investments, and level of living. We also explore the potential effects of these economic factors on social welfare and mental tranquilly. In order to achieve this, we created an extensive questionnaire and solicited replies from people in Bangalore who belonged to various demographic groups and professional backgrounds. Based on 423 gathered replies, the survey includes employees from a range of Bangalore-based businesses and industries. Following the development and testing of hypotheses, it was determined that an individual's position and type of job are significant factors influencing their well-being during the COVID-19 pandemic. This study illuminates the complex difficulties that ordinary citizens in cosmopolitan cities encountered during the epidemic and offers insightful information to decision-makers in government and local government.

Chapter 2 deep dives There could be an increase in fatalistic shifts in economic growth as a result of climate change. In the days ahead, the resources at hand will not be sufficient to meet the needs of the populace if the current pattern of trash distribution, eradication, and treatment persists. Natural resource availability has changed as a result of climate change. Treating all living things with respect and care is crucial for sustainable growth. Preserving the planet's vitality and preventing the depletion of its precious natural resources are more pressing concerns. Climate change has been made possible by the haphazard shift in individual attitudes, resource usage, and environmental practices. As a result, there may be less rainfall in all of India's regional zones; on the other hand, certain arid regions may also experience unexpected downpours. Certain renewable resources are scarce in our economy, but others—such as solar, tidal, and wind energy—automatically refill and renew themselves quite quickly. However, there are now too many obstacles caused by humans in many regions of the world when it comes to using water as a source. Since our economy depends heavily on non-renewable resources (such as coal, natural gas, petroleum products, etc.) and these resources are becoming increasingly scarce owing to population growth, a good "National Policy" for their preservation is still needed. The first step forward in this context was the introduction of the "National Natural Resources Management System (NNRMS)" scheme, and funding distribution for studies was also repeatedly updated. The greatest moment to take use of AI is now; the only thing left to worry about for the National Natural Interest is raising citizen awareness through public-private partnership. Regardless of educational background, it's critical to understand the worth of every resource in the economy and how human activity contributes to resource scarcity. One of the factors contributing to the implications of climate change is human behaviour in society. Being climate-responsible means going much beyond simply leaving trash cans outside for pickup. Recycling and generating less garbage have become vital components of our daily routine when we put more time and attention into them. Ultimately, this essay will be more theoretical in nature, discussing the causes of climate change, the availability of resources, government initiatives through different programmes, social responsibility, and the significance of going green. This post will also make it easier for everyone to understand the current state of our resource availability and the confident support they can give the economy by removing bad causes.

Chapter 3 intends nowadays, most people agree that climate change is a global emergency that has a significant impact on everyone's well-being, including mental health. Climate change has an impact on many mental health conditions, both directly and indirectly. Particularly in vulnerable groups, immediate psychological discomfort, encompassing anxiety, depression, and post-traumatic stress disorder, is one of the most direct consequences of climate change on mental health. The sense of helplessness in the face of continuous environmental change intensifies the feelings of sadness and bereavement associated with climate change-related incidents. Not to be overlooked are the indirect impacts of climate change-related migration and displacement, which lead to stress related to acculturation and the breakdown of social support systems. Changes in climate patterns can lead to economic insecurity, which can completely upend a person's life and finances. This may lead to financial strain and job loss, both of which have a strong correlation with mental health problems.

In Chapter 4, Catastrophes brought on by climate change are expected to drive human migration above current levels, as evidenced by their negative consequences on both physical and mental health outcomes. In the context of climate refugees—those who are compelled to relocate due to the effects of climate change—this chapter explores the complex relationship between resilience and mental health issues. It also looks at the various facets of resilience that people encounter when dealing with the psychological fallout from losing their homes. It looks at the existing legal systems that deal with climate refugees, trying to identify any gaps or inconsistencies that could be adding to the increased vulnerability that this specific population faces. This study intends to shed insight on the relationship between resilience, mental health, and legal complexity in order to impact policy deliberations and promote a more comprehensive strategy to addressing the well-being of climate refugees in the face of growing environmental realities.

The complex relationships between food security, mental health, social stability, and climate change are examined in this chapter. Important discoveries show that food security is disrupted by climate change, which affects populations and areas differently and affects availability, access, utilisation, and stability. Both direct and indirect influences on mental health can result in emotional effects, stress, and trauma. Social instability and health concerns, such as migration and violence resulting from resource constraint, are exacerbated by climate change. Specialised methods are required for vulnerable populations because they encounter particular obstacles. To gain a deeper understanding of the intricate connections between mental health and climate change, research must be improved. To address the implications of climate change for resilience in social stability and mental health, a comprehensive strategy combining public health, policy, and international collaboration is necessary. Proactive adaptation techniques, which include vulnerable populations, are crucial in mitigating harmful impacts on well-being, even though mitigation is still essential in this Chapter 5.

The global consensus is that sustainable products are necessary to save the environment. However, a variety of factors have led to a minor increase in ambient temperature in many urban areas these days. The widespread use of computers to operate all commercial applications is one of the causes. From the "Requirements Analysis" phase through the "Software Maintenance" phase, CASE (Computer Aided Software Engineering) tools are utilised to construct such systems. The Software Testing phase of the SDLC (Software Development Life Cycle) uses the most energy and computer resources in order to provide customers with high-quality software products in Chapter 6.

In Chapter 7, Urgent action is necessary in light of the growing concerns that climate change poses to ecological sustainability, food security, and human health, particularly for disadvantaged and marginalised people worldwide. With an emphasis on food and health outcomes, this paper examines the urgent need for programmes that both correspond with the Sustainable Development Goals (SDGs) and mitigate severe climate impacts. Acknowledging the complex and multifaceted effects of climate change on these essential facets of human welfare, there is an increasing demand for a comprehensive, scientifically grounded strategy. In order to effectively address climate change, this study explores the transdisciplinary practice of agroecology and positions it as a vital link between science, practical applications, and policy. The study looks at the importance of agroecology and how it might help communities become more resilient to climate change by creating sustainable food systems. The suggested strategy promotes a harmonic union of local knowledge and scientific understanding to alleviate access disparities through agroecology and lessen the effects of climate change on communities with limited development. To create more egalitarian, sustainable, and healthful food systems is the main objective. The effects of climate change are becoming more noticeable in India's changing environment, and they pose serious risks to mental health, food security, and social stability. Using technology to protect these essential components of wellbeing becomes increasingly important as the country deals with the effects of climate change. This essay examines the transformative potential of creative technical solutions in tackling India's particular climate change challenges.

Chapter 8 takes Social media usage is now an essential component of students' academic growth. Obtaining up-to-date knowledge and current information is the primary goal of using social media, and it is widely acknowledged that these platforms offer the most effective means of doing so. The purpose of this study is to present the impact of social media on students' intellectual growth. Survey research methodology was used in this study to gather primary data from 468 students. Social media's relatively low R-squared explanation of its influence was examined using the ANOVA test to determine the significant value.

Chapter 9 throws In this paper, we examine cyberstalking, drug trafficking on the dark web, and online retail crime to highlight the various issues that come up when attempting to criminalise, prevent, and regulate these acts on social media. It would be too difficult and outside the purview of this research to investigate cybercrime while recharging DTH incidents and their newsworthiness in traditional media in detail. Rather, the process of agenda-building is the main focus of this essay, with particular attention to social media, the influence of the social environment, and the ability of "ordinary" people to influence agenda definition. We also look into the intended audience's reach, the potential for a "trial by social media," and the role of social media in regard to issues related to pre-crime, monitoring, and prevention. Lastly, we challenge social scientists and criminologists to approach the challenges posed by new technologies with greater courage and rigour.

In Chapter 10, The statement "Without mental health, there can be neither health nor sustainable development" emphasises how closely related mental health and the SDGs are. As a result, reaching the SDGs is a step towards attaining mental wellness. One major factor that has a favourable influence on communities' mental health and general well-being is tourism. Bamboo is referred to as "Green Gold" because of its ability to sequester carbon and other environmentally beneficial qualities. These qualities help to lower carbon emissions and achieve the SDGs. The researcher has developed a model that attempts to address climate change issues, advance well-being, and accomplish the SDGs.

Chapter 11 focuses Intricate With its extensive effects on the environment and society, climate change has become one of the most important issues of our day. In addition to its evident effects on the environment, it has led to a crisis that is frequently disregarded: the severe psychological harm it causes to both individuals and communities. This chapter explores the complex web of feelings, concerns, and discomfort that climate change engenders, emphasising eco-anxiety and the psychological effects of climate change. The complexity of the problem is examined, along with its causes, effects, and possible remedies, highlighting how urgent it is to deal with the psychological costs associated with the climatic catastrophe. This chapter offers readers a greater knowledge of the emotional problems faced by climate change as well as a road map for navigating these feelings. It does this by providing a thorough analysis of eco-anxiety and climate-related misery. By addressing the psychological effects of the climate catastrophe, it highlights the significance of comprehensive solutions that close the gap between environmental and mental health, demonstrating the interdependence of combating climate change and promoting mental health in general.

Chapter 12 deep dives This chapter explores the nuanced ethical issues at play when mental health and climate change collide. The psychological well-being of individuals and groups is being affected by changes in the global environment, and successful and culturally appropriate solutions must be guided by ethical considerations. The chapter examines the moral obligations that communities, legislators, and mental health practitioners have when tackling issues related to mental health and climate change. It manages the challenges of including Indigenous viewpoints, encouraging international cooperation, and striking a balance between short-term mental health requirements and long-term sustainability objectives. Ethical quandaries and case studies provide practical insights into the difficulties faced by practitioners, emphasising the need for thoughtful decision-making procedures. The call to action that follows the chapter's conclusion emphasises a group commitment to moral values that place a premium on justice, inclusivity, and the integration of environmental stewardship and mental health. By examining and resolving these ethical issues, this chapter seeks to improve the ethical basis, equity, and resilience of addressing the substantial effects of climate change on mental health and well-being.

Chapter 13 intends moving overseas is already a tough choice that causes anxiety for most people. Sadly, the effects of severe climate change are driving people to move around the world, which is painful. Sea level rise, natural disasters, soil erosion, and agricultural destruction are some of the effects. These impacts start an irreversible chain reaction. It begins with environmental problems, which develop into biological problems that affect people's bodies, minds, and social systems both directly and indirectly. Furthermore, this type of displacement isn't reducing at all; rather, it will only increase if nothing is done to stop it. This essay seeks to present correct facts regarding the effects of migration brought on by climate change, emphasising the psychological effects while skipping over the social and physical ones. The instances and exemplars in the article cover a wide age range, from young children to adults and old individuals. It also covers isolated occurrences that are unique to a portion of the community, such as situations where women are primarily affected and situations where the psychological impacts of displacement have not been as severe as those felt by the average immigrant. The information was gathered from official websites, publications from reputable sources, and a range of academic journals. The information is based on data images and original world scenarios from different parts of the world in addition to the author's words. We have been able to determine which patterns appear to be more common than others as well as their specific scale impacts thanks to these situations. In addition, we now know a number of strategies and tactics to mitigate the negative effects of climate change on mental health.

In Chapter 14, This analysis shows how the effects of climate change pose a threat to the financial stability of the economy. This article aims to address the lack of knowledge on the extent of the effects that climate change poses to the financial industry, as research and statistics on the subject are still considered to be superficial and insufficient. By looking into several economic models, including GEM-E3-ISL and GINFORS, one can learn a great deal about the various economic sectors and systems affected by climate change. The examination of production functions and climate-related damages also highlights the direct and indirect effects on GDP, supply-side, and demand-side economic mechanisms. The chapter's conclusion highlights the significant impact on the economy, its sectors, and its structure. There is proof that climate change is affecting a wide range of sectors, chief among them being the economy. Even after we have computed the possible outcomes, data indicates that the effects of climate change are expected to be substantially more significant than we can currently comprehend.

Climate anguish, another name for eco-anxiety, is the term used to describe a persistent fear of environmental catastrophes brought on by realising how unavoidably climate change would harm the ecosystem. Raising worries for one's own survival as well as that of future generations is another aspect of eco-anxiety. One component of eco-anxiety is feeling helpless or uneasy about how one's personal life and the environment are affected by changes in the climate. Because eco-anxiety, also known as climate distress, is a common occurrence worldwide and has a major detrimental effect on people's functioning, research on its effects on mental health is essential. Eco-anxiety affects social behaviour, cognition, productivity, and behavioural responses in both positive and negative ways. These effects ultimately affect an individual's mental health and general well-being. By taking action on climate change, one can lessen its harmful effects in this Chapter 15.

Each chapter encapsulates a unique perspective, contributing to a comprehensive understanding of the intricate relationship between climate change and mental well-being.

IN SUMMARY

At the end of this long voyage delving into the complex relationship between mental health and climate change, our curated reference book serves as a reminder of the relationship that exists between environmental changes and human health. We have traversed the complicated terrain where climate-induced concerns meet with the delicate fabric of our mental and social spheres through the varied lenses of eminent experts and thinkers.

Each chapter has shed light on the various aspects of this crucial nexus, from the effects of the COVID-19 epidemic on Bangalore's cosmopolitan population to the far-reaching effects of migration and economic instability brought on by climate change. The topics covered in the talks ranged from the short-term psychological anguish brought on by climate-related incidents to the long-term economic effects and moral issues that direct our actions.

Climate change has accentuated a number of issues, including eco-anxiety, climate hardship, migration, food security, technology sustainability, and the moral compass directing solutions. These issues have given rise to a holistic picture of the opportunities and challenges that lie ahead.

The need for action—action that cuts across boundaries, disciplines, and ideologies—resounds throughout these pages. It's an appeal for international collaboration, careful policy adjustments, community-based projects, moral reflection, and technical advancements.

This conclusion isn't meant to be a farewell, but rather a call to action for all parties involved, including legislators, researchers, educators, business experts, and others. It's an exhortation to put knowledge into practice, close the knowledge gap between theory and practice, and forge a route towards wellbeing and resilience in the face of environmental disruptions.

We, the editors, would like to express our appreciation to all of the contributors whose knowledge and wisdom have enhanced this compilation. We also want to express our sincere gratitude to everyone of our readers for their participation, dedication, and will to take on the tasks described in these pages.

I hope that this edited reference book serves as more than simply a knowledge base; may it also serve as a catalyst for change, a road map for a time when balancing environmental sustainability with mental health is a key component of our group's goals.

Debabrata Samanta
Rochester Institute of Technology, Kosovo

Muskan Garg
Mayo Clinic, USA

Acknowledgment

We express our great pleasure, sincere thanks, and gratitude to the people who significantly helped, contributed and supported to the completion of this book titled *Impact of Climate Change on Mental Health and Well-Being*. Our sincere thanks to Fr. Benny Thomas, Professor, Department of Computer Science and Engineering, CHRIST (Deemed to be University), Bengaluru, Karnataka India, and Siddhartha Bhattacharyya,, Principal, Rajnagar Mahavidyalaya, Rajnagar, Birbhum, India for their continuous support, advice and cordial guidance from the beginning to the completion of this book.

We would also like to express our honest appreciation to my colleagues at Department of Computational Information Technology, Rochester Institute of Technology, Kosovo, Europe and Mayo Clinic, Rochester, Minnesota for their guidance and support.

We also thank all the authors who have contributed some chapters to this book. This book would not have been possible without their contribution.

We are also very thankful to the reviewers for reviewing the book chapters. This book would not have been possible without their continuous support and commitment towards completing the chapters' review on time.

To complete this book, all the publishing team members at IGI Global extended their king cooperation, timely response, expert comments, guidance, and I am very thankful to them.

Finally, we sincerely express our special and heartfelt respect, gratitude, and gratefulness to my family members and parents for their endless support and blessings.

Chapter 1
Cosmopolitan Living During the COVID-19 Era:
Navigating Challenges and Uniting Communities – An Impact Analysis

Lakshmi
https://orcid.org/0000-0002-6278-714X
School of Commerce Studies, Jain University, India

H. R. Swapna
Independent Researcher, India

Sheetal V. Hukkeri
School of Commerce Studies, Jain University, India

Digvijay Pandey
https://orcid.org/0000-0003-0353-174X
Department of Technical Education, IET Lucknow, India

Mukundan Appadurai Paramashivan
https://orcid.org/0009-0009-5608-4788
Aligarh Muslim University and Champions Group, India

ABSTRACT

The COVID-19 pandemic has dealt significant blows to economies worldwide, affecting people from all walks of life, including those residing in cosmopolitan cities. Bangalore, a city known for its diverse population comprising individuals from various ethnic backgrounds, languages, cultures, and customs, exemplifies cosmopolitan living. This study aims to examine the repercussions of the COVID-19 pandemic on the common people, or 'commoners', in Bangalore. The research explores the pandemic's impact on the well-being of individuals living in a cosmopolitan city, with a focus on factors such as earnings, investments, and standard of living. The authors also delve into how these economic aspects may influence mental peace and social health.

DOI: 10.4018/979-8-3693-2177-5.ch001

INTRODUCTION

Pandemics gives great blows to the economy. Covid-19 has affected the common people who even live in cosmopolitan city. People from various part of the world with different ethnic backgrounds, different languages, cultures and customs live in Bangalore. Hence we can address Bangalore as a Cosmopolitan city. "Bangalore is a very cosmopolitan city(times of India.com). varied people live in Bangalore. A study of the effect of covid-19 on commoner was undertaken to analyze the impact of Covid-19 on common people. Across the different demographic factors, the responses of respondents are analyzed to know the extent of impact of covid-19 on commoner.

The study was done to know the impact of covid-19 on the wellbeing of a commoner of a cosmopolitan city. The wellbeing of the commoner depends on earnings, investment and standard of living and this further may impact the mental peace and social health of people. A well-structured questionnaire was developed to know the effect of this pandemic on respondent from different demographic background and occupy different positions in the organizations. Widespread outbreaks of infectious disease, such as COVID-19, are associated with psychological distress and symptoms of mental illness (Bao et al., 2020). Psychiatrists across the world should be aware of these manifestations, their correlates, and strategies to manage them that encompass both the needs of specific populations (Yang et al., 2020) and the precautionary measures necessary to contain the spread of COVID-19 (Liu et al., 2020a). They should also be aware of lacunae in the existing literature, which may need to be filled in over time through more widespread clinical experience and research.

OBJECTIVES

1. To study the impact of covid-19 on the wellbeing of commoner of a cosmopolitan city.
2. To analyze the effect of covid-19 on Earnings, Investments and Standard of Living.
3. To study the impact of these factors on the mental peace and social health of the people.

Scope of the Study

The respondents are selected on the basis of their Job description and demographic background. The study was conducted by taking the perception of respondents who reside in Bangalore, India. The parameters developed for the analysis would study the impact of Covid-19 on commoner in Bangalore.

METHODOLOGY

- **Research Design:** Empirical research was undertaken to analyze the perceptions of respondents regarding the aforementioned topic.
- **Sampling Technique:** Purposive sampling was employed to gather data tailored to the research objectives, targeting individuals with specific characteristics.
- **Sample Unit:** The study focused on salaried employees occupying various positions across different organizations.
- **Sample Size:** A total of 423 participants were interviewed, representing diverse demographic backgrounds. The sample included 36 gray-collar workers, 130 blue-collar workers, 96 pink-collar workers, 98 red-collar workers, and 63 white-collar workers.
- **Primary Data Source:** Data collection was carried out through a structured questionnaire, featuring different parameters. Respondents were asked to evaluate the service quality of their banking institutions using a Five-Point Scale, where the highest score (5) indicated 'Strongly Agree,' and the lowest score (1) corresponded to 'Strongly Disagree,' alongside other objective measurement scales.
- **Secondary Data Source:** Supplementary information was gathered from various articles, journals, reports, websites, and e-books.
- **Data Analysis Tools:** The study took into consideration respondents from diverse demographic backgrounds and analyzed the data based on two demographic variables: Education and Gender. The analysis was driven by the research questions and objectives, underpinned by several hypotheses. To test these hypotheses, various statistical techniques were applied, including:
    - Two-way ANOVA
    - Correlational analysis
    - Regression analysis
    - Post Hoc Test
    - Multiple Response Analysis
- **Limitations of the Study:** Despite comprehensive coverage, the study is not without limitations. Although the pandemic affected a large population, only the perceptions of 423 respondents were analyzed. Additionally, time constraints constrained the selection of a larger sample size.

- **Data Analysis:** The data analysis process employed analytical and logical reasoning to extract insights from the collected data. The primary objective of this study was to assess the impact of COVID-19 on commoners residing in cosmopolitan cities. It incorporated both primary and secondary data sources. A well-structured questionnaire was designed, encompassing various parameters, each containing qualitative questions aimed at understanding respondents' perspectives on the topic at hand. Secondary data was utilized to provide a theoretical foundation for the study. The study population consisted of respondents working in diverse roles within organizations.

HYPOTHESES

H_0: There is no relationship between level of job and Mental wellbeing during covid period.

Interpretation: Since the p value is less than 0.05 we **reject** null hypothesis (Ho).

Table 1. Correlations

			Impact on Mental Peace	**Level of Job**
Spearman's rho	Impact On Mental Peace	Correlation Coefficient	1.000	-.464
		Sig. (2-tailed)		.000
		N	423	423
	Level of Job	Correlation Coefficient	-.464	1.000
		Sig. (2-tailed)	.000	
		N	423	423

Level of job is correlated with the mental peace. There is a significant relationship between the job level and the mental peace during the covid period. The people who occupy higher jobs have lesser mental stress. This is being described the negative correlation between the two. The extent of relationship is moderate. But the significant relationship prompts us not to accept the null hypothesis. Hence it can be said that there is a relationship between level of job and Mental wellbeing during covid period.

To know the extent of relationship between the type of job and impact on mental peace, regression anlaysis was applied. It is evident that the predictor variable mental peace has a sufficient impact on the dependent variable mental peace. The R2 =0.672 depicts that model equation explains 67.2% of the variance in mental

Table 2. Model summary[b]

Model	R	R Square	Adjusted R Square	Std. Error of the Estimate
1	.819[a]	.672	.671	.279

a. Predictors: (Constant), Level of Job
b. Dependent Variable: mental Peace

peace. Over all regression coefficient is statistically significant with p value <.05. hence the hypothesis is rejected by saying that there is a significant impact of the type of job on the mental peace of the respondents.

H_{01}: There is no difference in the different types of job on the amount of PF withdrawal.
H_{02}: There is no difference in the different age groups of respondents on the amount of PF withdrawal.
H_{03}: Type of Job occupied by the respondents and their different age groups has no effect on amount of PF withdrawn simultaneously.
H_1: Statistically there is a significant difference on the amount of PF withdrawal by the respondents from different age groups and having different types of jobs.

Interpretation: See below:

1. Since the sig. value is Less than 0.05, we reject null hypothesis (H_{01}).
2. Since the sig. value is More than 0.05, we are accepting null hypothesis (H_{02}).
3. Since the sig. value is more than 0.05, we are accepting null Hypothesis (H_{03}).

In the above table the effect of age and occupation on amount of PF withdrawal is shown separately as well as simultaneously.

The effect of age on the amount of PF withdrawal is analysed with the help of F ratio and associated P value. The Low F-value (0.564) and high p-value (0.689) of the parameter observed from the above table for the factor age of respondents, verifies that the null hypothesis with respect to age can be accepted at 5% level of significance. Therefore, it may be concluded that There is no difference in the average amount of PF withdrawal among the different age groups of respondents.

The high F-value (4.259) and low P-value (0.002) of the parameter observed, verifies that null hypothesis with respect to the different types of jobs occupied by the respondents, is to be rejected at 5% level of significance. Thus, it is concluded that type of Job has an effect on amount of PF withdrawal.

In order to know which group of respondents having different types of jobs, are significantly different from others, Post Hoc test is used. The multiple comparison table compares the one category of data with the rest of the categories selected for test. The mean difference column shows the mean difference in the amount of PF withdrawal. The

Table 3. Tests of between-subjects effects

Source	Type III Sum of Squares	df	Mean Square	F	Sig.
Corrected Model	384939358302.769[a]	19	20259966226.462	3.479	.000
Intercept	304390543826.986	1	304390543826.986	52.275	.000
age	13138059287.944	4	3284514821.986	.564	.689
Occupation	99192935371.464	4	24798233842.866	4.259	.002
age * Occupation	77474310247.094	11	7043119113.372	1.210	.278
Error	2346637001035.293	403	5822920598.102		
Total	3752400000000.000	423			
Corrected Total	2731576359338.062	422			

Dependent Variable: Amount of withdrawal from PF a/c
a. R Squared = .141 (Adjusted R Squared = .100)

amount of PF withdrawn of grey collar jobs is significantly different from pink- and blue-collar jobs since the p-value is less than 0.05. Likewise, the amount of PF withdrawal of white-collar and red-collar jobs is significantly different from blue- and pink-collar jobs.

From the homogeneous subsets it is observed that pink and blue-collar jobs make a homogeneous subset whereas the other 3 types of jobs form another subset. Hence it can be interpreted that pink- and blue-collar jobs are significantly different from other types of jobs and there is no significant difference in the amount of PF withdrawn of grey, white and red collar jobs.

The interaction effect of age and occupation shows the low F ratio (1.210) and high P-value (0.278). Since this value is higher than 0.05, the Null hypothesis is accepted. That is statistically there is no significant difference on the amount of PF withdrawal by the respondents from different age groups and possessing different types of jobs.

This pandemic has affected mental health of the respondents across the different age groups having different types of jobs. The different factors which affected the mental wellbeing of the respondents is shown the table. Majority of the respondents suffered from increased anxiety due to uncertain conditions prevailing. More than 55% of the respondents suffered from loss of appetite and Depressive dis-order. There are other factors like higher appetite and worsening chronic conditions which has disturbed the mental peace of the respondents.

FINDINGS AND CONCLUSION

The paper mainly focused on the impact of covid-19 on the respondents who hail from different demographic background. Hypothesis was developed to analyse the

Table 4. Post hoc test

Multiple Comparisons						
Scheffe						
(I) Type of job	(J) Type of job	Mean Difference (I-J)	Std. Error	Sig.	95% Confidence Interval	
					Lower Bound	Upper Bound
GCJ	BCJ	-68525.64*	14371.48	**.000**	-112999.1	-24052.17
	PCJ	-66979.17*	14913.18	**.001**	-113128.9	-20829.34
	RCJ	-21343.54	14871.62	.725	-67364.75	24677.67
	WCJ	-25198.41	15942.87	.645	-74534.65	24137.82
BCJ	GCJ	68525.64*	14371.48	**.000**	24052.17	112999.11
	PCJ	1546.47	10268.74	1.000	-30230.82	33323.76
	RCJ	47182.10*	10208.29	.000	15591.88	78772.32
	WCJ	43327.23*	11714.05	.009	7077.34	79577.11
PCJ	GCJ	66979.17*	14913.18	**.001**	20829.34	113128.99
	BCJ	-1546.47	10268.74	1.000	-33323.76	30230.82
	RCJ	45635.63*	10957.78	.002	11726.07	79545.19
	WCJ	41780.75*	12372.66	.024	3492.76	80068.75
RCJ	GCJ	21343.54	14871.62	.725	-24677.67	67364.75
	BCJ	-47182.10*	10208.29	.000	-78772.32	-15591.88
	PCJ	-45635.63*	10957.78	.002	-79545.19	-11726.07
	WCJ	-3854.88	12322.53	.999	-41987.75	34278.00
WCJ	GCJ	25198.41	15942.87	.645	-24137.82	74534.65
	BCJ	-43327.23*	11714.05	.009	-79577.11	-7077.34
	PCJ	-41780.75*	12372.66	.024	-80068.75	-3492.76
	RCJ	3854.88	12322.53	.999	-34278.00	41987.75

Dependent Variable: Amount of withdrawal from PF a/c

same. The statistical tests were applied to know the extent of relationship. It was found that there is a relationship between the level of job of respondents and the extent of mental peace they had during the covid period. Pandemics are deadly for any type of economy and India is no exception. There are different types of sectors in the economy and each sector is affected differently. There are many sectors which are showing a positive development post the lock down restrictions. But the phase of lockdown affected many sectors which were contact intensive. The respondents who work for these sectors had to face severe mental stress because of fear of job loss and because of financial stress. There was a severe job loss in contact intensive

Table 5. Post hoc tests (homogeneous sub sets)

Amount of Withdrawal From PF a/c			
Scheffe			
Type of Job	N	Subset	
		1	2
GCJ	36	4166.67	
RCJ	98	25510.20	
WCJ	63	29365.08	
PCJ	96		71145.83
BCJ	130		72692.31
Sig.		.437	1.000

Alpha = .05.

Table 6. Mental peace: Frequencies

		Responses		Percent of Cases
		N	Percent	
Mental Well Being During Covid Period[a]	Loss of appetite	195	20.9%	55.1%
	Higher appetite	134	14.4%	37.9%
	Worsening chronic conditions	103	11.1%	29.1%
	Increased anxiety	305	32.7%	86.2%
	Depressive dis- order	195	20.9%	55.1%
Total		932	100.0%	263.3%

a. Dichotomy group tabulated at value 1.

industries. On the other hand few sectors which were able to shift to WFH options were able to reduce the operational cost and the employees who work for such organizations and who draw a lucrative pay are experiencing a lesser mental strain. As the level of job increases the mental stress reduces. To analyse the extent of impact of type of job on the mental peace of the respondents, regression analysis was applied. It again showed that there is a sufficient impact of type of job on the mental peace of the respondents.

Most of the Blue collared and pink collared workers experienced a financial stress because of which they had to make a withdrawal from their PF account. Few of them had to withdraw for their lively hood. This further laid stress on mental peace. Hypothesis was developed to analyse the stress on those employees who are drawing a lesser pay. Anova test was applied for the same. It was found that the type of jobs has an influence on the amount of PF withdrawal. The age of the respondents was also studied on the amount of PF withdrawals, to know whether it has any influence on the later. It was found that age is not placing any impact on the amount of PF withdrawal. Even the interaction effect of age and the type of job also has no direct association with the amount of PF withdrawals. To know which type of job influence the amount of PF withdrawals more, Post hoc test with homogeneous sub tests were applied. It was found that Pink and Blue collared workers are different from the other set of workers. Hence it was interpreted that pink- and blue-collar jobs are significantly different from other types of jobs and there is no significant difference in the amount of PF withdrawn of grey, white and red collar jobs.

Multiple response analysis on the frequencies of the different covid-19 effects, it was found that majority of the respondents suffered from increased anxiety and next to that is depressive disorders as well as loss of appetite. There are other effects like higher appetite and worsening chronic conditions also.

Conclusion

Pandemics gives great blows to economies. Every sector was affected by this pandemic and during lockdown restrictions, majority of the respondents suffered from one or other physical or mental issues. The study was conducted to know how this pandemic affected the daily life of majority. It was found that most of the respondents who have not much paying jobs suffered mental and financial stress. The different types of jobs and their respective position and pay in the organization will influence the level of stress experienced by them. Different sectors are experiencing different types of recovery. Hence it is necessary to understand the impact of this pandemic on people representing different sectors in order to extend a supportive hand to reduce the extent of mental and financial stress.

Scope for Future Research

It is difficult to predict the type of recovery which will be experienced across the sectors in the country. There is an ample scope to study the impact of this pandemic on the economy from different angles. So many more researches on this topic will help economy to visualise the recovery of sectors and take corrective measures.

REFERENCES

Adhikari, S. P., Meng, S., Wu, Y.-J., Mao, Y.-P., Ye, R.-X., Wang, Q.-Z., Sun, C., Sylvia, S., Rozelle, S., Raat, H., & Zhou, H. (2020). Epidemiology, causes, clinical manifestation and diagnosis, prevention and control of coronavirus disease (COVID-19) during the early outbreak period: A scoping review. *Infectious Diseases of Poverty*, *17*(9), 29. doi:10.1186/s40249-020-00646-x PMID:32183901

Asmundson, G. J. G., & Taylor, S. (2020a). Coronaphobia: Fear and the 2019-nCoV outbreak. *Journal of Anxiety Disorders*, 70. PMID:32078967

Asmundson, G. J. G., & Taylor, S. (2020b). How health anxiety influences responses to viral outbreaks like COVID-19: What all decision-makers, health authorities, and health care professionals need to know. *Journal of Anxiety Disorders*, *71*, 102211. doi:10.1016/j.janxdis.2020.102211 PMID:32179380

Banerjee, D. (2020). The COVID-19 outbreak: Crucial role the psychiatrists can play. *Asian Journal of Psychiatry*, *50*, 51. doi:10.1016/j.ajp.2020.102014 PMID:32240958

Bao, Y., Sun, Y., Meng, S., Shi, J., & Lu, L. (2020). 2019-nCoV epidemic: Address mental health care to empower society. *Lancet*, *22*(395), e37–e38. doi:10.1016/S0140-6736(20)30309-3 PMID:32043982

Liu, S., Yang, L., Zhang, C., Xiang, Y. T., Liu, Z., Hu, S., & Zhang, B. (2020). Online mental health services in China during the COVID-19 outbreak. *The Lancet. Psychiatry*, *7*(4), e17–e18. doi:10.1016/S2215-0366(20)30077-8 PMID:32085841

Yang, Y., Li, W., Zhang, Q., Zhang, L., Cheung, T., & Xiang, Y.-T. (2020). Mental health services for older adults in China during the COVID-19 outbreak. *The Lancet. Psychiatry*, *7*(4), e19. doi:10.1016/S2215-0366(20)30079-1 PMID:32085843

Chapter 2
How a Country Like India May Turn a Threat Into a Bundle of Emerging Market Opportunities Under the "Go Green Initiative": Climatic Change May Lead to Resource Scarcity

Vinodini Kanniappan Jayachandran
https://orcid.org/0000-0001-9413-1199
SRM Institute of Science and Technology, Chennai, India

Alpa Srivastava
SRM Institute of Science and Technology, Chennai, India

ABSTRACT

The changes in the climate may lead to a rise in fatalistic changes in economic growth. If the same pattern of consumption, distribution, eradication, or treatment of waste is continued, the available resources will fall short of meeting the needs of the people in the coming days. Climate change has brought a change in the availability of natural resources. For sustainable development, it is more important to respect and care for all the forms of resources of life. Due to this, there is a potential drop in rainfall in the population in all regional parts of India. Contrarily, sudden rainfall in some dry areas is also witnessed. Finally, this chapter will be more conceptual and will bring about the reasons for climate change, resource availability, government interventions through various schemes, social responsibility, and the importance of going go-green. This chapter will further help all individuals to know the present status of resource availability and the support that they can provide to the economy from poor causes with greater confidence.

DOI: 10.4018/979-8-3693-2177-5.ch002

INTRODUCTION

In India, is water scarcity a myth or fact? People in India, understanding the water scarcity, consider it more precious than any Jewel. The report says that there is an extreme level of water scarcity, and India still looks upon the water needs for the monsoon. According to the report of NITI Ayog, the dependency on monsoons has increased the problem for the government to save and provide water to the population. The Asian Development Research Institute (ADRI) report says that there is a change in temperature due to global warming and which is a highly having impact on hydrologic processes in regional water bodies in India. The consequence of climate change has become more apparent and pivotal in finding innovative solutions towards the mitigation of shortage in natural resources and also creating breakthroughs that can provide optimum utilization of the available natural resources. According to the World Bank Report, there has been a gradual increase in the temperature by 2 Degrees Celsius to 4 Degrees Celsius which will affect global warming, mainly in the areas like agriculture, water bodies, and economics of the coastal areas.

In India, the preservation and availability of natural resources have become quite challenging because there is a rapid growth in population, urbanization, growth in the manufacturing sector, and dependence on fossil fuels. This challenge has come up with a change in climate conditions in different states in India. These points affirm how they are contributing to the paucity of resources and aggravating the effects of climate change. Some ofthe mitigations resulting from the "Go Green Initiative" the Government started initiatives to provide incentives for clean energy projects like the International Solar Alliance. India is targeting to enhance a worldwide leader in renewable energy (Pariatamby Agamuthu et al., n.d.). The farmers have started to build some success stories by adopting sustainable agriculture and water management techniques. This has also helped them to explore and innovate organic products and helped the country to increase the export of agricultural products in international markets. The country like India, which have started to face a shortage of natural resources and raw materials, have now made a great effort by not just collecting and keeping the waste but have started to recycle the waste (Devika Kannan et al. 2024). The country has started to identify waste management not as a menace but it has started to consider as a suitable chance to create an opportunity for profit-making business ideas—the dynamics with regard to this waste management from discarding, segregation, reduction, recycling and reuse (Neethu C et al. 2023). The Government of India has introduced providing certification for green building and urban greening projects by focusing on the economic benefits like constructing efficient buildings and good public transportation systems with urban parks. Urbanization and smart city projects have to go under the Green Umbrella.

The government is in ready mode to provide incentives and subsidies and has got regulations which will encourage green initiative projects in this growing economy, whereas this will be pivotal and crucial for the country to have international collaboration, financial funding support and technology exchange in accomplishing the vision of "Go Green Initiative".

India has the potential to transform the threat of natural resource scarcity due to climatic changes into an ocean of innovative opportunities in the open market (Saran Aadhar et al. 2019).The initial step towards it has inspired and motivated the stakeholders from civil society, government and industries to cooperate and unite in order to grab timely, the need gap offered under this green umbrella. This will result in holistic development and growth of the nation as well as the entities. This article focuses on the support engine "Go Green Initiatives" which will reshape the economic and natural health of the nation by converting the resource scarcity threat into emerging opportunities in the long run.

REASONS FOR CLIMATE CHANGE

Climate change occurs not only because of human and human activity but also the nature has its process for climate change. There are internal factors, such as plate tectonics, ocean circulation, carbon cycle and feedback mechanics, and external factors, like orbital change, volcanic activity, and solar variability. This natural process cannot be stopped, and it has occurred throughout Earth's history. However, the current concern is the increased rate of change due to human activity. The leading causes of this climate change are greenhouse gas emissions (GHGs), deforestation, changes in land use due to industrial processes, farming practices, growing population and rapid urbanisation. The accumulation of gases like methane (CH_4), carbon dioxide (CO_2), hydrofluorocarbons (HFCs), nitrous oxide (N_2O), and sulphur hexafluoride (SF_6) these gases capture heat by increasing the temperature of the earth's surfaces which creates global warming. The deforestation, changes in land use and changes in farming practices are acting as catalysts in trapping the heat in the environment because of inadequate trees (Micheal Shafer et al., n.d.). Due to population growth, people shift from rural to urban areas in search of survival prospects, which has led to urbanization and industrial growth. Further adding to the problem of resource scarcity, lapse in identifying better usage of existing resources, inefficient utilization of resources, untreated disposal of waste, recycling practices and unethical production processes.

CHANGE IN RESOURCE AVAILABILITY

The central aspect of resource availability that causes the pretentious climate crisis is a scarcity of food production, water, energy resources, forest and biodiversity, coastal resources, minerals and extractive industries. The changes in the resources have a greater impact on the ecosystem. The increase in demand due to population propulsion has not been met due to inadequate infrastructure. The reduction in the water table due to the over-exploitation of groundwater and changing precipitation patterns has increased the gap that already exists. Due to the shift in weather patterns, the frequency of extreme weather conditions, like floods, droughts, and cyclones, has increased, which has further reduced the availability of resources (Haider Ai et al.2019). The flora and fauna of the forest are affected due to the increase in cultivated land and commercial land. India is a developing economy that is rapidly growing there by increasing the energy consumption of fossil fuels. The emission of pollutants into the atmosphere by industries, and vehicles, burning of trash and release of affluents in the air have severely affected the air quality, which has resulted in health issues in living beings.

IMPACT

The ecological system, earth's system and living beings are adversely affected due to climate change, such as an increase in temperature, changes in the patterns of precipitation, the rise in sea level, severe changes in weather conditions, and ecosystem disorder. Insecurity in food products has increased, and the farmers are in a vulnerable situation against cultivating good quality, high-yielding crops (S K Dash et al. 2011). The rise and spread of infectious diseases and respiratory disorders brought on by poor air quality and heat-relatedness are a few of the health hazards that climate change increases. This dispersion of diseases that is such as mosquitoes, has an effect by changing the temperature and rainfall patterns, which has got an effect on the prevalence of illness, including dengue, malaria, covid etc. The economic repercussions adversely affected all the three sectors of the economy, such as primary, secondary and tertiary (Jew Das et al. 2022).

The Global initiatives to cut the emission of greenhouse gas, create sustainable lifestyles and increase resilience in susceptible places to combat climate change to guarantee a sustainable future for both present and future generations. The below table No. 1 represents the state-wise distribution of districts' rainfall on a five scale. The phase of rainfall shifts in months and not in seasons. There is a shift in place of rainfall and the record of rain is also in quantum.

Figure 1. Shows the climate change cause and evidence

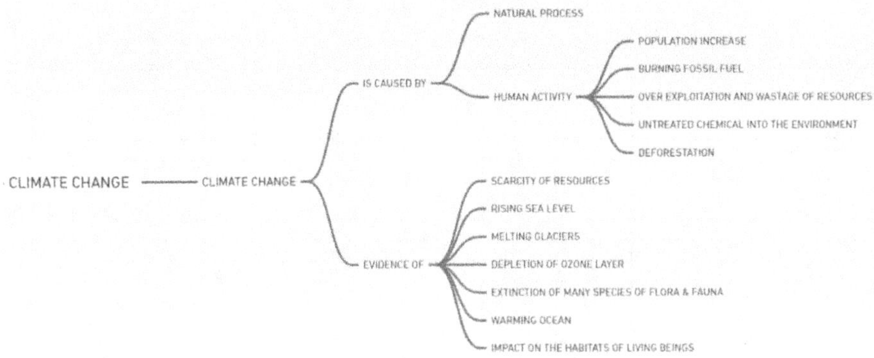

PARTICIPATION OF A CITIZEN

It is very important for a citizen to understand the change caused by climate and how the resources are becoming less adequate. On that note, the citizen is becoming self-educated and well aware nowadays of the changes caused in the environment. Government and technological advancement play a high role in helping the citizens to know about the changes on time. The citizens have understood that we humans are one of the sources for eradicating the natural resources, creating scarcity and polluting the environment with landfills. This understanding of a citizen has made them initiate a greater step to protect our environment. Schools, colleges, governments, NGOs, and the common man have stepped into planting trees and afforestation in understanding that it will decrease the level of carbon dioxide in the atmosphere. There are recycling centres which educate them about the benefits of recycling the waste (Aninda Bhattacharya et al.2023). This helps the business personnel to produce products by handling the waste and utilization of wastes in business through the recycling process. At all levels, the Indian education system also promotes the importance of protecting the environment by highlighting the issues in the environment and how to incorporate some sustainable programmes. NGOs, educational institutions, corporates, scouts and guides, NCC, NSS camps, and governments also organize clean-up activities like removing plastic bags, papers, bottles, and other waste in public places like schools, parks, roads, etc.

The manufacturers are in the process of producing more eco-friendly products and initiate themselves in go-green activities. In recent days the common man is more supportive in using eco-friendly products. The reduction in usage of plastic bags is an encouragement to use reusable ones. Burning of unwanted waste in the open-air is harmful to the environment. The release from burning leads to air

Table1. The state-wise distribution of districts' rainfall and its changing pattern

STATEWISE DISTRIBUTION OF DISTRICTS
WITH LARGE EXCESS,EXCESS,NORMAL,DEFICIENT,LARGE DEFICIENT AND NO RAINFALL

SN O	STATES	Period:01-06-2023 To 23-06-2023							
		LARGE EXCESS	EXCESS	NORMAL	DEFICIE NT	LARGE DEFICIE NT	NO RAIN	NO DATA	TOTAL
1	ANDAMAN & NICOBAR (UT)	0	2	1	0	0	0	0	3
2	ARUNACHAL PRADESH	0	3	7	4	1	1	0	16
3	ASSAM	4	9	10	4	0	0	0	27
4	MEGHALAYA	3	4	0	3	1	0	0	11
5	NAGALAND	1	0	6	4	0	0	0	11
6	MANIPUR	1	1	1	4	2	0	0	9
7	MIZORAM	0	0	3	5	0	0	0	8
8	TRIPURA	0	1	3	4	0	0	0	8
9	SIKKIM	0	2	1	1	0	0	0	4
10	WEST BENGAL	1	1	1	5	11	0	0	19
11	ODISHA	0	0	1	12	17	0	0	30
12	JHARKHAND	0	0	0	6	18	0	0	24
13	BIHAR	0	0	1	2	34	1	0	38
14	UTTAR PRADESH	8	3	6	14	40	4	0	75
15	UTTARAKHAND	0	1	4	2	6	0	0	13
16	HARYANA	2	3	5	5	7	0	0	22
17	CHANDIGARH (UT)	0	0	0	1	0	0	0	1
18	DELHI (UT)	1	1	3	3	1	0	0	9
19	PUNJAB	6	5	4	5	2	0	0	22
20	HIMACHAL PRADESH	1	2	7	1	1	0	0	12
21	JAMMU & KASHMIR (UT)	2	2	7	7	0	1	1	20
22	MADHYA PRADESH	4	1	4	3	40	0	0	52
23	GUJARAT	9	4	4	5	11	0	0	33
24	DADRA & NAGAR HAVELI AND DAMAN & DIU (UT)	0	0	0	1	2	0	0	3
25	GOA	0	0	0	0	2	0	0	2
26	MAHARASHTRA	0	0	0	0	35	1	0	36
27	CHHATTISGARH	0	0	0	3	24	0	0	27
28	ANDHRA PRADESH	0	1	1	18	6	0	0	26
29	TELANGANA	0	0	0	9	24	0	0	33
30	TAMIL NADU	12	8	9	5	4	0	0	38
31	PUDUCHERRY (UT)	1	1	0	0	2	0	0	4
32	KARNATAKA	0	0	8	7	16	0	0	31
33	KERALA	0	0	0	5	9	0	0	14
34	LAKSHADWEEP (UT)	0	0	0	1	0	0	0	1
35	LADAKH (UT)	1	0	0	0	1	0	0	2
36	RAJASTHAN	19	3	7	3	1	0	0	33
	TOTAL	76	58	104	152	318	8	1	717
CATEGORYWISE DISTRIBUTION OF DISTRICTS OUT OF 716 FOR WHICH DATA RECEIVED		11%	8%	15%	21%	44%	1%		

Note: "The Rainfall values are rounded off up to one place of decimal."
Source: www.indiaenvironmentportal.org.in

pollution, toxic emissions, carbon dioxide and methane. Some of the hazardous chemicals are having a high impact on causing health effects and condemnation of the environment. The citizens are even educated on the burning of a variety of waste and how illegal dumping has got an impact on global climate change. A zero-waste lifestyle has also been promoted and followed by a few, which have highly encouraged the business personnel to reduce and eliminate waste in business operations. Constructing an eco-friendly building with green architecture is also a trending practice. The Government have also made a clear understanding and the importance of saving rainwater. The protection against rainwater systems has been followed in all construction as a mandatory practice. In recent days, the preference for buying and consuming organic products has increased, and citizens have started to show high interest in harvest plants, terrace gardening, and indoor planting. This participation and initiatives taken by citizens after understanding the negative causes in the environment like climate change, resource scarcity, negative health effects etc.

GOVERNMENT INTERVENTIONS

The nation's beauty is its diverse population, and it has also become a challenge for the nation related to the availability of resources for all. India, being a developing economy in its economic upswing, is facing a lot of environmental pressure and challenges. The various efforts to address the problem of scarcity of resources by the government of India fall in various areas like water resource management – Jal Shakti Abhiyan, Jal Jeevan Mission, Rainwater Harvesting, Watershed Development, Water Conservation campaigns, Drip Irrigation and Sprinkler Irrigation in agriculture, Ground Water Management, River Basin Management. The main source of the Indian economy is the Agricultural sector; the Indian government has improved water consumption through the existing agricultural practices by granting subsidies for structured irrigation systems, organic farming, and assisting crop diversification, which will lead to soil health refinement. The promotion of renewable energy by government drives such as the National Solar Mission and National Wind Energy and efforts are put to tap tidal energy.

The government of India has taken up afforestation programs and forest conservation for flora and fauna. With the help of the waste management process, some industries develop innovative ideas for manufacturing bricks, footwear, home decor, environmental decor like idols in parks, roadside, and government construction of roads. The individual, society and government have taken initiatives and steps to recycle and segregate the waste to make recycling easier by splitting into various forms of bins like brown (organic waste), yellow (hazardous waste), orange (solid waste), red(general household cannot go for recycle), green (recyclable waste), and blue (liquid waste). In energy efficiency, the county is becoming efficient in producing electricity with the help of solar, wind and water energy. This is worked indifferent sectors in the nation to reduce the consumption of the available resources and the discharge of greenhouse gases. For example, The Ministry of New and Renewable Energy (MNRE) has taken steps to include all the solar modules and solar cells in e-waste management. The schemes have made all the producers to be accountable for storage as well as the recycling of the waste.

ADAPTATION AND MITIGATION

India is combating internationally by taking the initial step in the Paris Agreement for reducing the emission intensity, decreasing fossil fuel-based energy and expanding the forest cover where India is only be G20 country among the nations.

The government takes a few steps to keep a check on resource scarcity, and they are also promoting a few measures in all the sectors for consistent and continuous resource management.

Table 2. Key Commitments for India's NDC (Nationally Determined Contributions) to the Paris Agreement. The Goal Outlines between 2021 and 2023

Key Commitments for India's NDC (Nationally Determined Contributions) to the Paris Agreement: The Goal Outlines Between 2021 and 2023	
S.No	**Goals**
1	LIFE- Lifestyle for the environment
2	Adopting environment-friendly and clean work mechanism
3	To reduce the intensity of emissions by 45% by 2030
4	To build a carbon sink for 2.5 to 3 billion tonnes of CO_2
5	To adapt the climate change in endangered sectors like agriculture, Himalayan region, water resources, coastal regions, disaster management and health.
6	Cutting-edge of climate technology with international collaborations for internal diffusion.

SOCIAL NON-COMPLIANCE

The pain points for Social non-compliance on the part of people at large it is important to address: the increasing domestic usage of energy, accumulation of food which is more than required, usage of fuel during daily transport, increasing consumption of water due to the number of ways it is used and the quantity along with it, too much of landfills is accumulated by the people in the environment. Accumulation of plastic, e-waste, food waste, environmental cleanliness, medical waste, and cutting down trees for construction to increase production usage of chemical fertilisers and pesticides have ruined the soil fertility and resulted in the overall imbalance of the ecosystem. Due to all these practices of human beings, the other species are becoming vulnerable along with us to the changing effects of climate and resource scarcity.

CONCLUSION

Climate change is a natural process; it has occurred many times in the history of the earth, causing temperature changes, changes in weather patterns, sea level changes, changes in the ecosystem, agricultural and food security, cultural and societal impact, ocean circulation, human settlements and global and regional impacts. Focusing on the anthropology of the societal members is important to diminish and redesign the causes of climate change. To have the sustainable availability of resources and mitigate the impact of climate change, we need to adapt and learn the measures to prevent the cause. Some of the measures related to mitigation can include energy efficiency, reduced deforestation, eco-friendly technology innovation, transition to renewable energy, afforestation and reforestation, carbon pricing, and sustainable agriculture. Regarding adaption, we must develop insurance and financial mechanisms to help the communities, biodiversity conservation, early warning systems, community engagement, and natural resource management. It is important to reduce raw material extraction by following the four R's (Reducing, Reusing, Recycling and Repurposing) of resources. Every citizen in the country has his responsibility to not cause harm to the environment and society. Getting educated and aware of the cause of pollution, population, burning of fossil fuels, and prevention is more important. The usage of vehicles, appliances, and products must be eco-friendly and biodegradable. The creation of monitoring systems for the availability of resources pattern of consumption to provide information for better decision making. Hence, it is significant to address climate change and resource availability, which leads at a national level to policy changes, social-level behaviour of individuals, industrial-level eco-friendly technological innovation and cooperation at the global level.

REFERENCES

Aadhar, S., & Mishra, V. (2019, November 6). A Sustainable rise in the area and population affected by dryness in South Asia under 1.5 Degree Celsius and 2.5 Degree Celsius Warmer Worlds, Environmental Research Letter. *IOP Science*, *14*(11). Advance online publication. doi:10.1088/1748-9326/ab4862

Ai, H., Modi, P., & Mishra, V. (2019, September). Increased Flood Risk In Indian Sub-Continent Under the Warming Climate, Weather and Climate Extremes. *Weather and Climate Extremes*, *25*, 100212. doi:10.1016/j.wace.2019.100212

All India Monthly rainfall Series. (n.d.). https://tinyurl.com/r9bfuzfu

Bhattacharya, Thomas, Soni, Roy, Sarangi, & Kanawade. (2023). Opposite Trends in Heat Waves and Cold Waves Over India. Journal of Earth System Science, (67).

BioOne VISTA. (n.d.). *Climate Change*. https://tinyurl.com/988as3u7

Central Pollution Control Board (CPCB). (n.d.). *Environmental Protection Act & Rules*. https://cpcb.nic.in/

CNBC TV 18. (n.d.). https://tinyurl.com/ye26xhh2

Das & Umamahesh. (2022). Heat Wave Magnitude over India under Changing Climate: Projections from CMIP5 and CMIP6 experiments. *International Journal of Climatology*. doi:10.1002/joc.7246

Dash & Mamgain. (2011). Changes in the Frequency of Different Categories of Temperature Extremes in India. *AMS Journal*, 1842 – 1858. doi:10.1016/j.wace.2019.100212

Devika, Khademolgorani, Janatyan, & Alavi. (2024). *Smart Waste Management 4.0*. doi:10.1016/j.wasman.2023.08.041

Government of India. (n.d.). https://tinyurl.com/m8cxvhtn

India Government. (n.d.). *Environment and Forest*. https://tinyurl.com/2s5ch437

Indian Council of Forestry Research and Education. (n.d.). https://icfre.gov.in/

Modi Yojanaye, P. M. (n.d.). *Jal Jeevan Mission (Rural) Scheme*. https://tinyurl.com/axmrtx2s

NAB. (n.d.). https://tinyurl.com/4sb36wy9

National Biodiversity Authority (NBA). (n.d.). *Know the Biological Diversity Act 2002 and the rules 2004, Information on Biodiversity*. https://tinyurl.com/5n7tb2ve

National Disaster Management Authority (NDMA). (n.d.). *Natural*. https://ndma.gov.in/

Neethu & Ramesh. (2023). *Projected Changes in Heat Wave Characteristics Over India*. Springer.

Pariatam. (n.d.). Waster Management Developments in the last five decades. *The Journal of Sustainable Circular Economy*. doi:10.1177/0734242X231199938

Pradhan Mantri Krishi Sinchayee Yojana (PMKSY). (n.d.). *About PMKSY*. https://pmksy.gov.in/

Saur Energy International. (2023). https://tinyurl.com/5b2hj7mz

Shafer. (n.d.). *Climate Change Primer, Warm Heart World Wide.* https://tinyurl.com/mvpus57u

Swachh Bharat Activities. (n.d.). swachhbharat.mygov.in

The Central Ground Water Board (CGWB). (n.d.). http://cgwb.gov.in/bhujal-news-0

The Ministry of New and Renewable Energy. (n.d.). *Annual Reports from 2016-2023.* https://mnre.gov.in/en/annual-report/

The WB IBRD. (n.d.). *IDA press release.* https://tinyurl.com/bp6y2au6

Water conservation under MGNREGA. (n.d.). https://tinyurl.com/yx6tcw72

Chapter 3
Climate Mindscape:
Navigating the Vital Link Between Climate Change and Mental Well-Being

Jyoti Khandelwal
University of Engineering and Management, Jaipur, India

Amrita Raj
University of Engineering and Management, Jaipur, India

ABSTRACT

It is now widely acknowledged that climate change is a worldwide catastrophe with far-reaching effects on human well-being, including mental health. Numerous mental health issues are influenced by climate change, both directly and indirectly. One of the most direct effects of climate change on mental health is the immediate psychological distress, which includes anxiety, depression, and post-traumatic stress disorder, especially in vulnerable communities. A sense of powerlessness in the face of ongoing environmental change exacerbates the melancholy and loss brought on by climate change-related events. The indirect effects of migration and displacement brought on by climate change are also noteworthy, as they result in acculturation stress and the dissolution of social support networks. A person's life and economy can be upended by economic insecurity brought on by shifting climatic patterns. This can result in job loss and financial stress, both of which are strongly associated with mental health issues.

INTRODUCTION

Addressing climate change stands out as one of the paramount challenges confronting the world today, necessitating thorough examination and thoughtful consideration (Goudet et al., 2023). Climate change is identical with a multitude of adverse

DOI: 10.4018/979-8-3693-2177-5.ch003

events, including but not limited to heat waves, elevated temperatures, deforestation, glacier melting, floods, tornadoes, hurricanes, droughts, fires, and desertification. The consequences of climate change extend beyond environmental adjustment and have a significant impact on society and biology. The intricate nature of these climate-related challenges has raised growing concerns among scientists due to their complexity, both in direct and indirect consequences. This worry has led the scientific community to look into the many ramifications, from the susceptibility of biological entities to the general susceptibility of societies.

The literature related to mental illness catch the attention of researcher in this area. The latest technologies can help expert to understand the actual metal health. The complexity of mind not depend on the single thing here the climate direct and indirect impact are also force researcher to work in this field (Kerz et al., 2017). The objective of this chapter is to understand the impact of climate change in human mind, and suggest how the latest technologies can help to improve the human life in changing climate.

Furthermore, not every aspect of society is equally affected by climate change (Li, C. et al, 2022).]. Geographic exposure and vulnerability, coupled with limited access to resources, information, and protection, emphasize the disparities in the impacts on mental health that distinct communities must contend with. Due to the intricate connection between climatic occurrences and mental health issues, new words such as ecopsychology, ecoanxiety, ecoguilt, ecobereavement, solastalgia, and biospheric concern have been developed (Palinkas et al., 2020).

This chapter attempts to navigate the unexplored area of how climate change affects mental health by dissecting the intricacies, illuminating the various temporal dimensions, and emphasizing the differences between various population groups. The goal of this chapter is to add to the growing body of knowledge regarding the relationship between mental health and climate change by exploring the body of existing literature and introducing new vocabulary. The general flowchart in figure 1, illustrates the rational progression of investigating the effects of climate change on mental health (Desai et al., 2021).

LITERATURE REVIEW

According to the (Eisenack et al., 2014), the notion of barriers in adapting to climate change is becoming more and more popular, signifying challenges in both planning and execution. Current research identifies common obstacles but FALLS short of providing thorough justifications and fresh perspectives. The authors advocate for actor-centered, comparative research methods that concentrate on interdependencies. They stress the importance of conducting exploratory research to comprehend the

Figure 1. Logical flow of exploring the impact of climate change on mental health

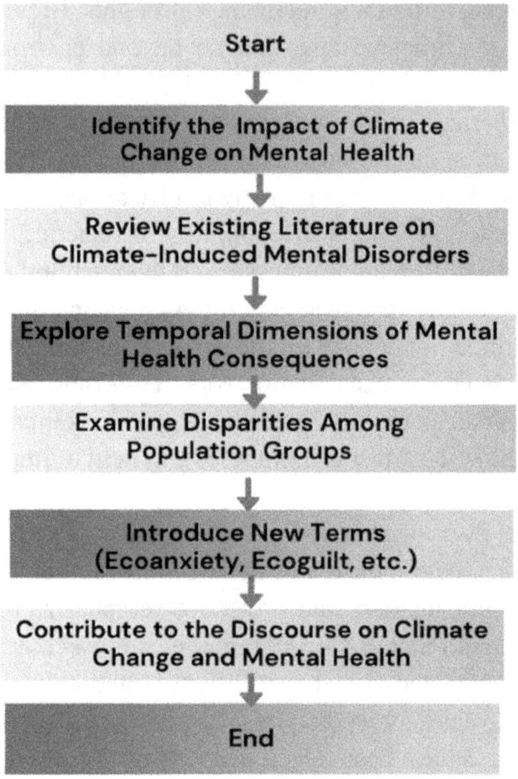

origins, relationships, and alterations in barriers over time. In order to transfer knowledge and make informed decisions in adaptation planning, comparative, actor-centered, and time-sensitive research is considered essential. The Perspective calls for a change in research focus from descriptive to explanatory in order to improve our knowledge of obstacles and enable better-informed climate adaptation strategies.

As it is discussed in (Padhy et al., 2015), which talks about the wide-ranging effects on mental and physical health, mentioning correlations between rising temperatures and higher incidences of violence and violent suicides. The relationship between farmer suicides and climate-related conditions like droughts is examined, as well as the wider consequences for mental health, stress, and disorders. This study also emphasizes that the necessity of considering mental health when the growing mitigation strategies which discusses the possibilities of the psychological consequences of the population mobility that brought by climate change. In the end, it advocates for more proactive measures to address the projected mental health issues linked to climate change as well as for more accessibility to mental health treatments.

According to the (Charlson et al., 2021), the global research priorities of the WHO, this scoping review looks at 120 original studies (2001-2020) on the relationship between mental health and climate change. Most studies, mostly cross-sectional and quantitative, concentrate on evaluating the risks to mental health that climate change poses, especially for high-income nations. Exposure to climate-related factors, such as heat, humidity, and wildfires, has been associated with increased mortality for people with pre-existing conditions, worsened mental health, and psychological distress. On the other hand, a dearth of studies emphasizes the importance of evidence-based mitigation and adaptation strategies, addressing other research priorities. Despite existing limitations, the overwhelming body of research indicates that there is a negative relationship between mental health and climate change. This suggests that future studies should prioritize interventions to protect mental health in the face of climate change and produce strong, methodologically sound findings.

In light of this (Hayes et al., 2018), which describes the current and anticipated risks that climate change poses to mental health, focusing on three main areas: the difficulty of linking particular climate change risks to mental health outcomes, the increasing impact on mental health that disproportionately affects marginalized communities, and the necessity of coordinated interventions based on active hope. The study highlights the connection between mental health and climate change and ends with recommendations for urgent efforts. Even though sometimes it can be challenging to draw the direct links, in order to fighting climate change is an idea to be essential for resilience and mental health activities. By addressing the effects of climate change on mental health in the twenty-first century, the need of efforts for global commitments and coordinated, while it is also called for revolutionary action were rooted in health justice.

On the basis of (Lawrance et al., 2022), by this thorough analysis, it looks the relationship between mental health and climate change which highlights the developing dangers and that give impact on the health of the entire community. According to the report, climate change is a risk multiplier that disrupts the circumstances which is critical to mental health, that directly effects on the mental health and inequitably impact to the poor. Some hazardous climate change such as extreme weather and rising temperatures intensifies the pre-existing stressors, which gives a negative impact on the mental health. This review highlights how to organize quickly the measures which is based on the climate justice for the need of active hopes. Understanding the serious implications, it also emphasizes how bold global leadership may lead to transformative action and a better, more sustainable future while highlighting the connections between mental health, climate change, and societal well-being.

As it is discussed in (Hwong et al., 2022), this scoping review of climate change and mental health research reveals a rapidly growing field with 56 eligible articles identified between 2000 and 2020 which highlights the fast expanding field of research on climate change and mental health. The main approaches used to obtain data on self-reported mental health consequences, hospital admissions, and suicide rates connected to climate events were surveys, administrative health records, and vital statistics. The study encourages the partnership among the mental health professionals and the data scientists which highlights the necessity of linking the databases of the mental health outcome to weather data for causal inference. In order to comprehend the danger of psychological disorder as a result from both acute and chronic climate change, future research needs to be developed towards the sophisticated approaches. By addressing the potential long-term repercussions, research and public health attention are giving priority to the relationship between climate change and mental health.

MATERIALS AND METHODS

The effects of environmental changes on physiological well-being is meritorious o the attention that can take many more different forms (Corvalan et al., 2022). High temperatures, harsh weather conditions, and environmental disturbances is the reason of increase in psychological distress and emotional strain. The research shows a direct link between elevating temperatures and an increase in suicide rates, and emphasize the deleterious repercussion of climate change on mental health. Vulnerabilities are amplified, especially in disadvantaged people, which aggravates mental health symptoms. Heat waves and natural catastrophes are examples of climatic disruptions that act as pressures to aggravate pre-existing mental health conditions. The growing field of "eco-anxiety" explains the psychological toll that growing public awareness of climate change takes, impacting people even in situations in which they are not directly affected. Comprehensive research, early intervention, and addressing climate change are all necessary given the intricate relationships between mental health and climate change.

Events Cause Mental Health Changes

Extreme Weather Events and Mental Health

Elevating temperatures, severe weather, and disruptions to the environment Storms and floods are becoming more common and extreme weather occurrences due to climate change. Individuals who experience these may be exposed to potentially traumatic experiences, such as seeing someone seriously hurt or killed. Many will experience increased psychological anguish as a result, and a small percentage may experience more severe mental health issues like depression, substance abuse disorders, or post-traumatic stress disorder (PTSD). Because they might result in homelessness, unemployment, or insecurity over food and water, a few of the social and economic factors that influence mental health can also be impacted by extreme weather events. In consequence, this may have a negative impact on mental health (Williams et al, 2022). When harsh weather strikes, people, families, and communities often show exceptional amounts of resilience. A 2010 study found that the given right psychosocial support, many patients either fully recover or maintain good mental health. However, considering the growing frequency and intensity of extreme weather events, we are not sure if this sort of resilience can last over time. People, families, and communities frequently show very high levels of resilience when faced with extreme weather. According to 2010 research, most patients either fully recover or maintain excellent mental health with the correct psychosocial therapy. However, it is not clear that if this level of resilience can endure progressively more severe weather occurrences in the long run.

Heat and Mental Health

Rising global temperatures are the root cause of climate change, which can have a number of unfavorable effects on internal health. For example, during heatwaves, hospitalizations for mental illnesses and urgent mental health visits typically rise. The number of self-murders has also been demonstrated to rise during heatwaves, and this trend is expected to continue as temperatures rise, though there is conflicting evidence linking heat and self-murder. What connects the heat to ill internal health is still not well understood. One theory holds that high temperatures can exacerbate people's moods and make them feel more twisted and anxious, which exacerbates the signs of internal health issues. It has been demonstrated that the duration of heatwaves is correlated with additional negative sentiment expressed on the internet. Heat can also interfere with sleep, and insufficient sleep can exacerbate internal health issues. According to a massive study that examined ten billion sleep compliances, warmer nights are eroding mortality sleep inversely and encyclopedically, with the effect being three times greater for people living in lower-income countries.

Air Pollution and Mental Health

It has been shown that as temperatures rise and the climate changes, allergens and other pollutants are more concentrated in the air, resulting in a decline in air quality. There is mounting evidence that unhealthy air quality can have a negative impact on mental health, including depression and anxiety. A large US investigation found that among all Medicare-eligible individuals over 65, short-term exposure to pollutants (PM2.5 and NO2) was associated with an increased risk of acute hospital admission for psychiatric illnesses.

In a similar vein, individuals with psychosis or mood disorders in the UK have been found to seek mental health services more frequently when air pollution levels are greater.

Research has also demonstrated a link between childhood and teenage exposure to air pollution and the emergence of mental health issues in young adults. It has been proposed that this results from air pollution affecting how the central nervous system develops normally.

Infectious Disease and Mental Health

A significant contributing factor to the spread of infectious diseases like dengue, zika, and malaria in new parts of the world is climate change. The alterations linked to global warming can provide information about the persistence, multiplication, cornucopia, and dispersion of infections, vectors, and hosts. If global warming is not slowed down, many deadly conditions will have more opportunity to spread, endangering the health of millions of people. Continue reading to find out how contagious complaints and climate change are related, as well as what can be done to mitigate the effects. Internal and physical health are inextricably linked. Being hospitalized or having to deal with the long-term effects of a serious infection are just two of the negative effects of being exposed to advanced rates of contagious conditions on internal health. People who live in specific environments, such as neglected tropical environments, may also experience discrimination and demarcation. A growing number of people report experiencing anxiety related to climate change, as well as other related cerebral responses, and direct exposure to climate hazards. There is also a growing report of cerebral responses related to defying the prospect of climate change. Although there isn't a clinical consensus for these gests, there have been multiple attempts to label them with alternative terminology.

These include (among others):

- **Nostalgia** – the incapability of chancing solace in a familiar geography due to environmental declination.
- **Ecological grief** – the sense of loss arising from passing environmental declination.
- **Climate anxiety** – a feeling of anxiety in the face of climate change.

One large study of 10,000 children and youthful people in 10 countries set up that 45 of repliers said their passions about climate change negatively impacted their diurnal functioning. Another study set up that negative, climate- related feelings were associated with further symptoms of wakefulness and poor internal health. Some argue that these represent adaptive and formative cerebral responses to climate change, and that we should not pathologies them. There are arising substantiation that advanced situations of climate anxiety can be, in some cases, linked to advanced situations of pro-environmental behaviors. Figure 2 illustrates the climate hazards and its vulnerability and exposure.

Figure 2. Climate change and mental health

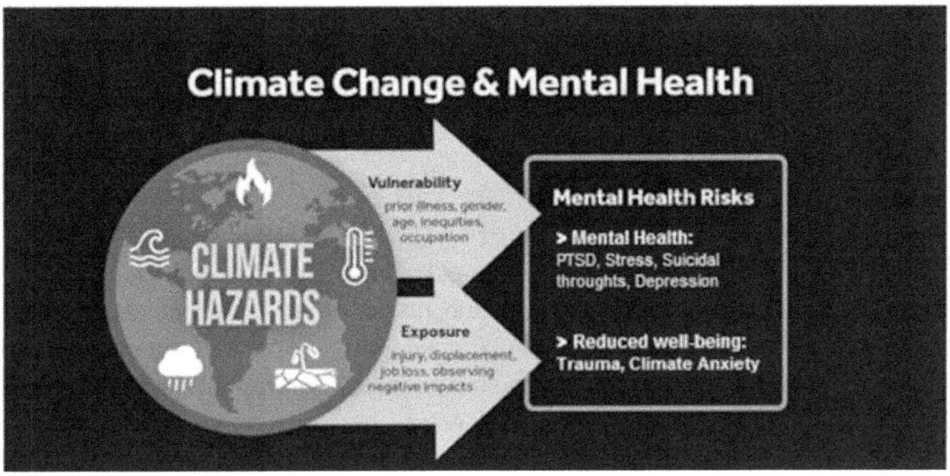

Mental Health Outcomes

Climate change's effect on mental health outcomes is a complex issue with wide-ranging effects. Numerous negative impacts on mental health are exacerbated by high temperatures, severe weather, and environmental disturbances. Studies demonstrate a direct relationship between rising suicide rates and climate change, highlighting the detrimental effects of environmental changes on mental health. People who are affected by both acute and long-term climate catastrophes report higher levels of worry and psychological discomfort. People who are already vulnerable, such as those with mental health issues, deal with additional difficulties like worsening symptoms and being more vulnerable to the negative consequences of climate change (Hepp et al., 2022). Additional factors that contribute to mental health costs include economic strain, living conditions upheaval, and the possibility of displacement as a result of climate-related catastrophes. The emerging concept of "eco-anxiety" demonstrates the psychological cost of being conscious of climate change, impacting people even in situations where they are not immediately affected. The identification of these intricate mental health consequences highlights the pressing requirement for comprehensive interventions, resilience-enhancing tactics, and worldwide endeavors to alleviate and adjust to the psychological obstacles presented by climate variability.

Symptom Scales and Screening Tools

When assessing the mental health, screening instruments and symptom scales which is a crucial tool. These tools are often used in clinical settings that offers measurable measures of the symptoms. Uniform measurement systems which improve the diagnostic precision and acetate the creation of efficient treatment regimens. These resources are crucial for promoting early intervention and enhancing mental health in general.

Hospital Admissions

It can be impacted by mental health results, demonstrating the connection between mental and physical health. Severe mental illness or crisis sufferers might need to be admitted to a hospital for evaluation, support, and care to the patient. It is necessary to admit in the hospital when the circumstances including severe mental disease symptoms, suicide thoughts, or acute psychotic episodes when it goes harsh day by day. The connection highlights how this is necessary to address the mental health holistically manner. Potentially decreasing the need for hospital admissions is possible with early intervention and community-based treatment of mental health.

Self-Harm

It means the self-destruction which results suicide and it is an upsetting the manifestation of mental health repercussions that occurs often. Individuals who are going through the emotional upheaval could turn to harming themselves by actions as they have the coping strategy. It can take more different forms, from cutting to other intentional harm to themselves, and it is frequently reflecting the underlying problems with mental health. In order to avoid self-harm, it is imperative to identify the causes of distress. That is why therapy is needed by early intervention, and community resources are necessary in decreasing the likelihood of self-destruction behaviors.

Burden of Disease

The outcome for mental health has a significant influence on the global burden of disease. Disorders such as schizophrenia, depression, and anxiety place a significant strain on both the individual and society. Mental health issues affect a person's ability to function, their quality of life, and their overall well-being, which has an impact on society and the economy. Treatment is just one aspect of addressing the stigma associated with mental health; other strategies include prevention, raising public awareness, and destigmatizing. Comprehensive mental health initiatives are critical to reducing the substantial burden and improving results for general health.

Scoping Review

It examines the complicated relationship between climate change and mental health is the subject of a scoping review, emphasize the increasing part of research highlighting the significant impact of environmental changes on mental health. The thorough review of the research highlights the various ways that climate change—which includes elevating temperatures, harsh weather conditions, and ecological disruptions—contributes to a range of mental health issues. From the direct effects of natural disasters, like trauma and post-traumatic stress disorders, to the subtle effects of prolonged exposure to environmental changes, like heightened anxiety and depression, the evidence shows how urgently the mental health dimensions of climate change need to be addressed. Furthermore, it appears that vulnerable populations—such as underprivileged and disadvantaged communities—are disproportionately affected, aggravating pre-existing health disparities. The scoping review highlights the need for comprehensive, interdisciplinary approaches that integrate mental health considerations into strategies for mitigation and adaptation to climate change as the field advances. This will help to foster resilience and promote holistic well-being in the face of this global challenge.

This review adhered to the PRISMA Extension shown in figure 3, for Scoping Reviews (PRISMA-ScR) (Tricco et al. 2018) and the Preferred Reporting Items for Systematic Reviews and Meta-Analyses (PRISMA) 2020 standards (Brennan et al. 2021). The following online databases were searched from January 1, 2001, to December 31, 2020, in order to find original research studies about mental health and climate change: PubMed, PsycINFO, EMBASE, CINAHL, Web of Science, and Scopus. Appendix A contains search strings for each database; the PubMed search string (Box 1) was modified for each database (Miller et al., 2023). Although we acknowledge that there are other pertinent and expansive concepts of mental health, such as "emotional wellbeing," we decided to limit our search keywords in order to create a boundary around an excessive number of references. By looking through the key review articles' reference lists, other studies were found. Two authors (SA and M.P.) separately retrieved data from studies that met the inclusion criteria into a Microsoft Excel sheet, and three authors (F.C., S.A., and M.P.) individually

Figure 3. PRISMA flow chart for scoping review on climate change and mental health

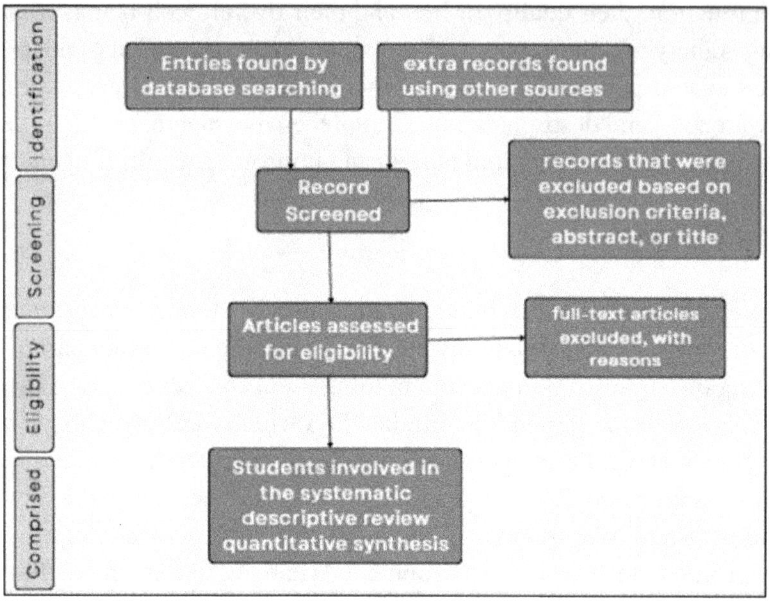

completed title/abstract and full-text screening using Endnote X9 (Desai et al., 2021). Throughout this procedure, these authors tried to reach a consensus.

Search Strategy

In order to pinpoint the main mechanisms by which climate change may impact mental health, a thorough analysis of the body of existing research and expert consultation served as the foundation for the search strategy to find the literature on this subject. The following terms related to mental health and climate change were included to identify pertinent literature on this topic based on this reading and expert consultation (Lawrance et al., 2022):

- **Climate change:** Temperature, heat, heatwave, warming, season, hot, flooding, floods, hurricane, drought, forest fire, thunderstorm, cyclone, tornado, bushfire, snowstorm, natural disaster, forced displacement, forced migration, climate refugee, climate migration;
- **Mental health:** Mental illness, wellbeing, suicide, psychosis, anxiety, depression, PTSD (posttraumatic stress disorder), eco-anxiety, ecoanxiety, psychoterratic, solastalgia, ecoparalysis, ecological grief, ecopsychology.

The search engines WebofScience, PsychINFO, Embase, PubMed, and Google Scholar were searched for terms relating to the effects of climate change on mental health and the full combination of search terms is provided in Supplementary Information.

Key Challenges of Attributing Climate Change to Mental Health

Four major factors make it difficult to assign blame for the relationship between climate change and mental health: first, there is a chance that common, transitory distress responses to unusual events will be pathologized, leading to an underdiagnosis of the effects of a changing climate on mental health; second, there is a wide range of possible outcomes from climate change that could affect mental health; third, there is significant variability in the timing of the effects of climate change on mental health, making causal relationships more difficult to establish; and fourth, attribution related to climate change and mental health is poorly understood due to the intricate relationship between mental health and other social determinants of health. Both pathologizing "normal" reactions to climate change and underdiagnosing its actual consequences on mental health are risks associated with this. Pathologizing "normal" reactions to disasters could make it difficult to distinguish between more serious, long-lasting mental health issues and mild transitional sadness or bereavement. Inaccurate prevalence estimates and distorted presumptions about the demand for mental health services can result from either over- or underestimation of the mental health outcomes linked to climate change. In the wake of Hurricane Katrina, Whaley also pointed out that in certain instances, medical personnel failed to evaluate pre-existing mental health conditions and consequently misdiagnosed patients as having a stress response or disaster trauma as a normal stress response when they had much more serious mental health problems related to the hurricane's effects that were not identified. It is vital to remember that climatic changes have the potential to aggravate or even cause pre-existing mental health disorders or issues.

Some mental health issues are sophisticated and multifaceted, requiring knowledge of the complex relationships between environmental influences and an individual's mental health. Some challenges associated with attributing climate change to the mental health are such as: *The complex causation, Time Lag and Cumulative Effects, Vulnerability and Resilience, Pre-existing Mental Health Conditions, Differential Impact on Population Groups, Geographical Variability, and Adaptation and Coping Mechanisms*. Pointing out these challenges is very difficult for developing effective interventions and policies.

Adaptation Measures

Some specific priority adaptation mechanisms that ought to be considered to support population-level mental health in a changing climate:

- **Policy responses:** Improving access and funding to mental health care is necessary.
- **Surveillance and monitoring:** Conducting epidemiological surveys following extreme weather events; and monitoring the emergency department that visits during heat waves and follows the extreme events.
- **Practice:** Stepped-care approach to mental health which is an application that frequently used in disaster the mental health for the support of different levels of interventions depending on the timing of the disaster and the level of distress.
- **Preparation and response:** Climate change adaptation/resilience planning in the mental health system.
- **Community-based interventions:** Plans for climate change resilience that address psychosocial wellbeing.
- **Special training for caregivers and first responders:** e.g. psychological first aid.

Preventions

If we want to avoid the worst effects of climate change on human health, we must shift away from our reliance on fossil fuels and toward clean, renewable energy, stop deforestation, and restore our natural habitats.

There are several co-benefits between various mitigation techniques for climate change and mental health. Given the connection between physical exercise and mental health, switching to more active forms of transportation, such walking or cycling, can have a favorable effect on mental health.

Likewise, expanding access to green areas can benefit mental health as well. Wellcome-funded studies have demonstrated that young people's moods can be instantly and momentarily lifted and their anxiety levels can be lowered by spending merely 15 minutes in a green environment like a park or forest.

Numerous effects of global warming are now irreversible, even if we manage to cut emissions and reach the goal of zero emissions worldwide by 2050. Therefore, we must also adjust to the ongoing effects of climate change if we hope to preserve population health in the future.

For many mental health conditions, there are numerous efficient evidence-based therapies available. Some have been tried with populations of migrants and refugees or in environments that may be especially relevant to climate change, including disasters. On the other hand, further study is required to determine the best ways to prevent and treat mental health issues in the context of climate change, as well as how to enhance mental health through early intervention.

Through his work on climate change and health, Welcome aspires to a world in which human health can flourish while catastrophic climatic collapse is prevented. This includes a day when, in light of climate change, nobody is prevented from achieving their goals by mental health issues. Figure 4, include give brief detail about prevention strategies by operationalizing these strategies, people and communities have the potential to enhance their preparedness.

Figure 4. Prevention strategies to operationalizing people and communities

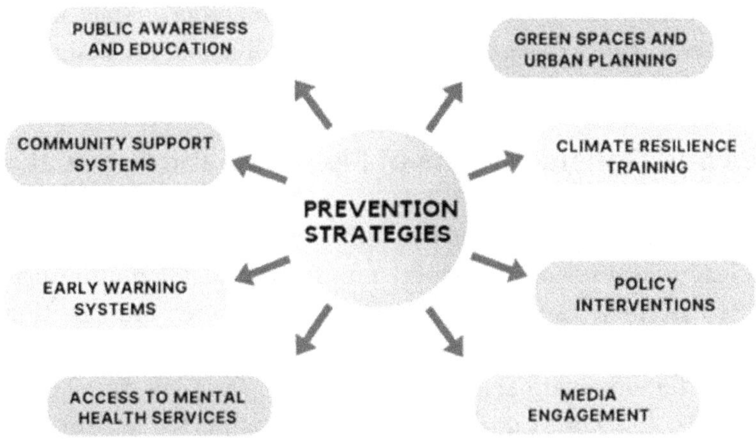

COMPUTER APPLICATION FOR MANAGING MENTAL HEALTH

You can install mental health apps on your computer, smartphone, or other smart device to help with emotional or mental well-being. They can help with anything from tracking mood and behavioral symptoms in people with anxiety, depression, and other mental health conditions to offering relaxation, mindfulness, and meditation exercises.

Mobile apps for mental health are made to assist users with common mental health issues, increasing mindfulness, and enhancing their emotional well-being.

Some apps for mental health give users access to psychiatric care or online talk therapy (Mange et. Al, 2023). Others might support you in starting a meditation practice, managing stress and anxiety, or getting better sleep. Usually intended to be used on a smartphone or tablet, mental health apps are available for users of iOS and/or Android.

Variety of Therapy Techniques Involve in Application

There exists a multitude of app formats, yet the pivotal factors lie in the quality and intentions driving these applications. Some apps encompass a diverse range of therapeutic techniques and mental health support, such as cognitive behavioral therapy (CBT), Dialectical Behavior Therapy (DBT), Exposure Therapy, Meditation, Mindfulness, Relaxation Exercises, and Stress Management. They work respectively for transformation of negative thinking patterns and behaviors, coping skills for handling challenging situations, exposes individuals to feared stimuli within a controlled, clearing the mind of distractions, promoting a state of heightened awareness, contribute to overall well-being, and Techniques designed to assist individuals in effectively coping with stress.

Features and Considerations of Mental Health Applications

When creating an application for the computer to manage mental health, user needs, privacy issues, and the moral ramifications of handling private data must all be carefully taken into account. Some features must be think before developing these kind of applications such as *User Verification and Profile Development* for safeguard sensitive data, *Mood Tracking* for visual depiction of mood trends, *Journaling and Reflection* for Prompts for self-reflection, *Goal Setting and Progress Tracking* for Feature for setting and tracking personal goals related to mental health, *Mental Health Resources* for information about coping mechanisms, stress reduction, and mental health education. Some other features are *Reminders and Notifications, Community Support, Therapist Connection, Crisis Support, Privacy and Security, Machine Learning for Personalized Insights, Feedback and Improvement, Gamification for Engagement, Customization, and Education and Awareness.*

Above all the features mention ethical considerations and user privacy should be prioritized throughout the development process. Consulting with mental health professionals and obtaining user feedback during development can enhance the effectiveness and user-friendliness of the application.

Selection of Accurate Application

All users are not always technically sound for them the selection of accurate application is really difficult work. Here some points are mention which help user to choose the correct app. The following points or factors are:

1. **Requirement and feature correlation:** Select an app whose features fit your requirements. For example, use an app that provides fitness exercise if you'd like assistance with body fitness stress relief.
2. **According to the user feedbacks:** Choose an app that has undergone testing and been shown to be successful in studies by the general users.
3. **User friendly:** The app should be look simple and easy to use.
4. **Platform independent:** Select an app that works with the smart devices you own. Certain apps are limited to particular operating systems and platforms.
5. **Authorization:** Make sure the app has privacy-protecting security features.

Pros and Cons of Using Mental Health Apps

There are many benefits to using mental health apps, such as:

1. Ease of use
2. Accessibility
3. Budget Friendly
4. Self-Placed

It's crucial to recognize the disadvantages of mental health apps, though. But there are some places where these app are still need improvement:

1. Can't able to replace human experts
2. Not have any supporting data from science
3. Privacy and security issues
4. No direct human interaction
5. Excessive reliance on technology

Examples of Mental Health and Therapy Apps

We selected the top apps based on their availability, techniques, user-friendliness, and focus on mental health. These five apps are a great place to start if you're looking to improve your mental health, based on these factors:

1. **Calm:** This application working for meditation to improve the human mind wellbeing. To make this app more helpful for the user it include instructional and medication guides. The new technical feature help human to achieving peaceful sleep which impact the overall growth of mind also the main goal of stress- free lifestyle is achieved. The most famous features are "Calm Body", and "Calm Kids".
2. **Headspace:** The motive of this app to provide guided classes for stress reduction. The working of this application based on activities which help to reduce the stress. The activities mainly focused on courses, music, audio experiences, sleep casts, and inspirational wake-up stories.
3. **Sanvello:** Designed with people experiencing stress, anxiety, or depression in mind, Sanvello incorporates exercises in cognitive behavioral therapy (CBT) with meditation therapy. Daily mood tracking is a feature of the platform that helps users identify trends and factors that affect their mental health.
4. **Happify:** This app, which promotes positive psychology, includes games, evidence-based mental health exercises, assessments, and progress tracking to improve mood and general wellbeing. Weekly challenges that are in line with their objectives—be they stress management, relationship enhancement, or job satisfaction—are sent to users.
5. **Bearable:** With its many features, this app makes it easy to track mood, mental health, physical symptoms, medication use, and much more. The app includes features like behavior trends, medication logs, activity logs, journaling, and reminders. Users can choose to synchronize important health data from their smart devices and set reminders. The app helps users better understand how different treatments affect their overall health by providing personalized insights and statistics in addition to tracking user progress.

PHYSIOLOGICAL CARRIERS TO CLIMATE CHANGE MITIGATION AND ADAPTATION

A major issue involving science, money, people, regulations, and morality is climate change. It affects the whole world, but we feel it in our neighborhoods. Even if we stop making pollution now, the Earth will keep getting warmer, and climate change will affect the kids and grandkids after us. The main thing causing this is a gas called carbon dioxide, and it stays in the air for a really long time. We hit a record in 2013 with more carbon dioxide than ever before, like millions of years ago. Because we can't stop everything right away, we need to do two important things:

- **Stop Making Pollution (Mitigation):** We have to reduce the things that make the Earth hotter, like cars and factories. This will help slow down how much the climate changes.
- **Get Ready for Changes (Adaptation):** Even if we try to stop pollution now, some changes are already happening. We need to be ready for those changes, like rising temperatures and extreme weather.

So, we have to work together to slow down climate change and be ready for the changes that are already on the way.

According to Figure 5, virtually all regions of the earth are experiencing rapid changes in temperature, precipitation, and/or other aspects of climate. Human-caused changes in the earth s climate are greatly altering ecological systems, with consequences for human well-being.

Conceptualizing Barriers to Climate Change Mitigation and Adaptation

Mitigation of climate change refers to actions taken to lessen the effects of human activity on the planet's climate by reducing or eliminating the emission of greenhouse gases. The brief overview present in Figure 6.

Figure 5. Conceptual diagram of mitigation, and adaptation options associated with climate change

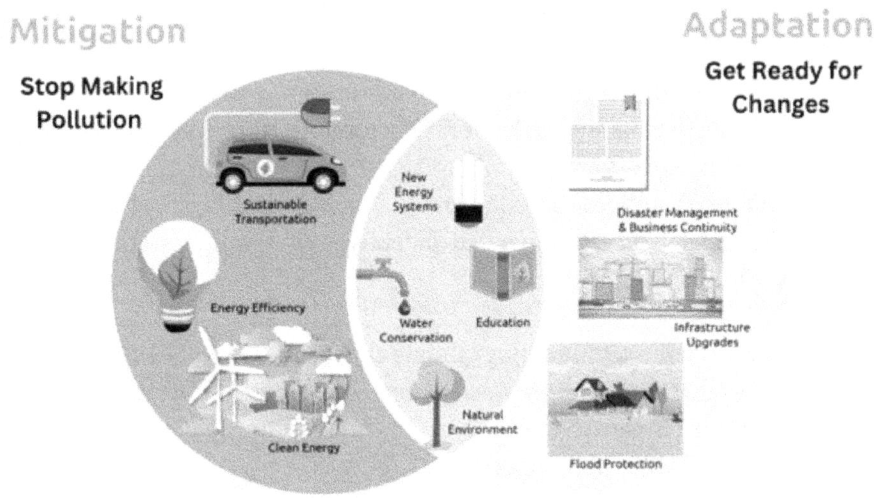

Figure 6. Conceptualizing barriers to climate change mitigation and adaptation

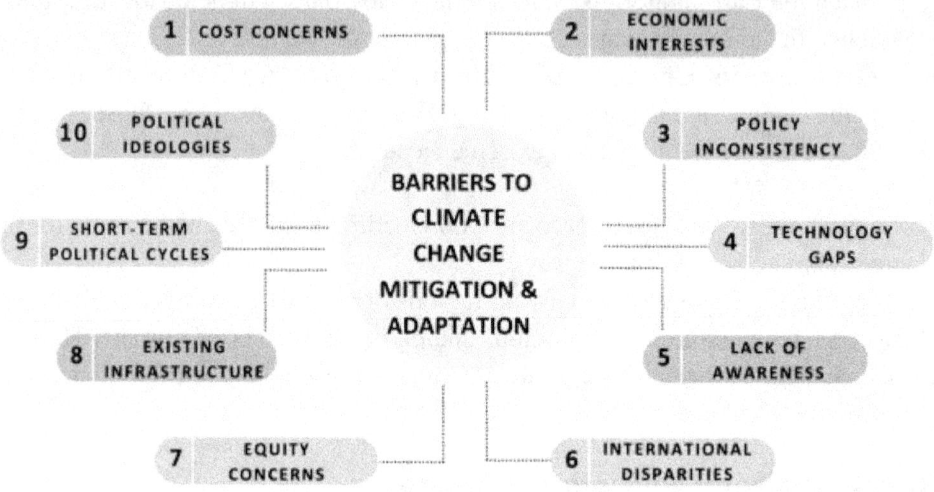

Nonetheless, a number of obstacles stand in the way of effective climate change mitigation. The following are some conceptualizations of obstacles to mitigating climate change:

1. **Economic Barriers:** The implementation of these technologies can be costly.
2. **Uncertainty:** Regulatory and Policy Difficulties in long-term planning can result from frequent modifications to climate policies.
3. **Effective Technologies:** The technological obstacles may be hampered by inadequate funding for research and development aimed at innovative climate solutions.
4. **Public Knowledge and Comprehension:** Apathy and inaction may result from a lack of social and behavioral Factors of the seriousness and urgency of climate change.
5. **Sustainable Practices:** If companies and individuals adopting budgetary obstacles do not receive financial incentives or rewards, mitigation efforts may be discouraged.

Some other obstacles to mitigating climate change are *Global Collaboration and Fairness Concerns, Infrastructure Restrictions, and Political Divide.*

To overcome the climate change and create secure future above method obstacles are necessary to follow by the society.

Conceptualizing Methods to Overcoming Barriers of Climate Change Mitigation and Adaptation

An all-encompassing and integrated strategy involving different stakeholders at the local, national, and international levels is needed to remove obstacles to climate change adaptation and mitigation. The brief overview present in figure 7.

Figure 7. Conceptualizing methods to overcoming barriers of climate change mitigation and adaptation

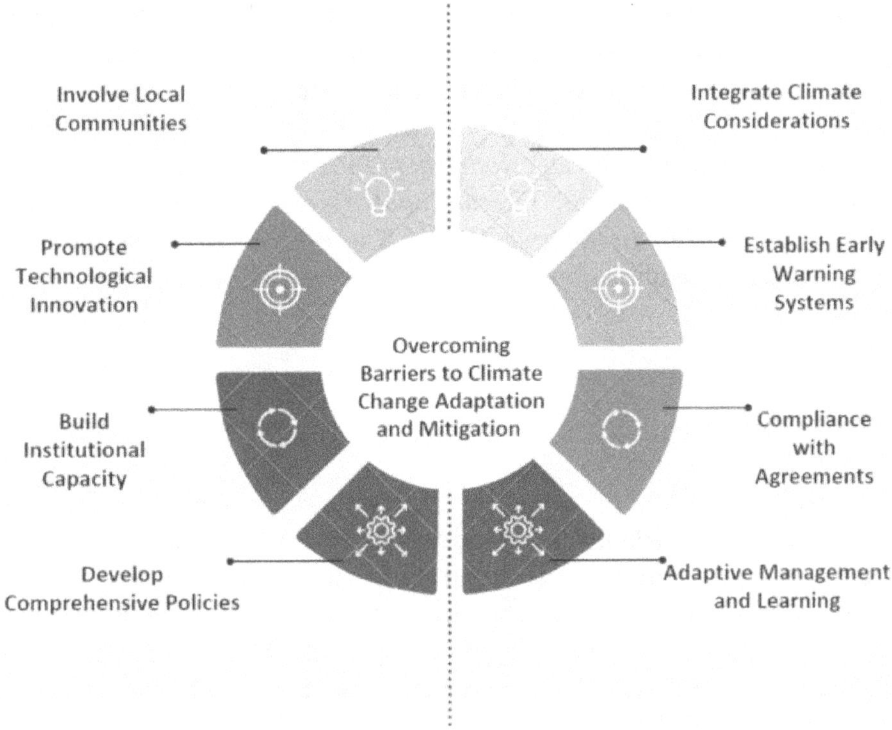

The following are ideas for how to get past these obstacles:

1. **Combined Governance and Policies:** To integrate climate change adaptation and mitigation into larger development strategies, governments should create and implement comprehensive policies. To guarantee efficient coordination, cooperation, and enforcement of climate-related policies, governance frameworks should be strengthened.

2. **Building Capacity and Educating:** Encourage programs promoting climate literacy and awareness to educate people about the significance of adaptation and mitigation, as well as communities and decision-makers. Make investments to enable institutions—such as non-governmental organizations and local governments—to organize and carry out climate-resilient projects.
3. **Innovation and Financial Assistance:** Offer vulnerable areas and communities financial assistance as well as incentives to adopt adaptation and mitigation strategies against climate change. Encourage the development of novel technologies that aid in sustainable practices and lower emissions by funding research and development.
4. **Social Inclusion and Community Engagement:** Taking into account their distinct needs and viewpoints, involve local communities in the planning and decision-making processes linked to climate change adaptation.

Some other obstacles are *Development of Infrastructure, Evaluation of Risk and Early Warning Systems, International Agreements and Collaboration, and Integrating Climate-Related Issues.* The effects of climate change by implementing a combination of these strategies and customizing them to particular contexts.

CONCLUSION

The investigation of how climate change affects mental health has covered a wide range of ground, including conceptualizing mitigation and adaptation plans and looking at everything from physiological carriers to causal events. In order to fully comprehend the complex relationship between climate change and mental health, this chapter addressed a number of important topics, including causes, effects, search techniques, obstacles, adaptations, preventative measures, and the use of technology in the form of mental health apps.

As a result of this chapter's illumination of the complex interactions between mental health and climate change, a thorough and integrated approach is imperative. Building resilient and adaptive societies requires addressing the mental health dimensions, which is not only morally required but also strategically necessary as we face the challenges posed by a changing climate.

REFERENCES

Brennan, S. E., & Munn, Z. (2021). Prisma 2020: A reporting guideline for the next generation of Systematic Reviews. *JBI Evidence Synthesis*, *19*(5), 906–908. doi:10.11124/JBIES-21-00112 PMID:33989266

Charlson, F., Ali, S., Benmarhnia, T., Pearl, M., Massazza, A., Augustinavicius, J., & Scott, J. G. (2021). Climate change and mental health: A scoping review. *International Journal of Environmental Research and Public Health*, *18*(9), 4486. doi:10.3390/ijerph18094486 PMID:33922573

Corvalan, C., Gray, B., Villalobos Prats, E., Sena, A., Hanna, F., & Campbell-Lendrum, D. (2022). Mental health and the global climate crisis. *Epidemiology and Psychiatric Sciences*, *31*, 31. doi:10.1017/S2045796022000361 PMID:36459133

Desai, Z., & Zhang, Y. (2021). *Climate Change and Women's Health: A Scoping Review*. Academic Press.

Eisenack, K., Moser, S. C., Hoffmann, E., Klein, R. J., Oberlack, C., Pechan, A., Rotter, M., & Termeer, C. J. A. M. (2014). Termeer, C. J: Explaining and overcoming barriers to climate change adaptation. *Nature Climate Change*, *4*(10), 867–872. doi:10.1038/nclimate2350

Goudet, J.-M., Arif, F. B., Owais, H., Ahmed, H. U., & Ridde, V. (2023). *Intersecting Vulnerabilities: Unraveling the Complex Relationship between Women*. Climate Change, and Mental Health in Bangladeshi Communities.

Hayes, K., Blashki, G., Wiseman, J., Burke, S., & Reifels, L. (2018). Climate change and mental health: Risks, impacts and priority actions. *International Journal of Mental Health Systems*, *12*(1), 28. doi:10.1186/s13033-018-0210-6 PMID:29881451

Hepp, J., Klein, S. A., Horsten, L. K., Urbild, J., & Lane, S. P. (2022). The Climate Change Distress and Impairment Scale: Introduction of the Measure and First Findings on pro-. *Environment and Behavior*.

Hwong, A. R., Wang, M., Khan, H., Chagwedera, D. N., Grzenda, A., Doty, B., Benton, T., Alpert, J., Clarke, D., & Compton, W. M. (2022). Compton, W. M.: Climate change and mental health research methods, gaps, and priorities: A scoping review. *The Lancet. Planetary Health*, *6*(3), e281–e291. doi:10.1016/S2542-5196(22)00012-2 PMID:35278392

Kerz, E., Wiechmann, D., & Riedel, F. B. (2017). Implicit learning in the crowd. *Studies in Second Language Acquisition*, *39*(4), 711–734. doi:10.1017/S027226311700002X

Lawrance, E. L., Thompson, R., Newberry Le Vay, J., Page, L., & Jennings, N. (2022). The impact of climate change on mental health and emotional wellbeing: A narrative review of current evidence, and its implications. *International Review of Psychiatry (Abingdon, England)*, *34*(5), 443–498. doi:10.1080/09540261.2022.2128725 PMID:36165756

Li, C., Lawrance, E. L., Morgan, G., Brown, R., Greaves, N., Krzanowski, J., Samuel, S., Guinto, R. R., & Belkin, G. (2022). Belkin, G.: The role of mental health professionals in the Climate Crisis: An urgent call to action. *International Review of Psychiatry (Abingdon, England)*, *34*(5), 563–570. doi:10.1080/09540261.2022.2097005 PMID:36165755

Mange, D., Vaidyanathan, M., Lanka, S. P., & Ahuja, S. (2023). Application for performing self care activities using flutter. *2023 2nd International Conference on Vision Towards Emerging Trends in Communication and Networking Technologies (ViTECoN).*

Miller, V. J., Maziarz, L., Wagner, J., Bell, J., & Burek, M. (2023). Nursing assistant turnover in nursing homes: A scoping review of the literature. *Geriatric Nursing*, *51*, 360–368. doi:10.1016/j.gerinurse.2023.03.027 PMID:37104907

Padhy, S., Sarkar, S., Panigrahi, M., & Paul, S. (2015). Mental health effects of climate change. *Indian Journal of Occupational and Environmental Medicine*, *19*(1), 3. doi:10.4103/0019-5278.156997 PMID:26023264

Palinkas, L. A., O'Donnell, M. L., Lau, W., & Wong, M. (2020). Strategies for delivering mental health services in response to global climate change: A narrative review. *International Journal of Environmental Research and Public Health*, *17*(22), 8562. doi:10.3390/ijerph17228562 PMID:33218141

Tricco, A. C., Lillie, E., Zarin, W., O'Brien, K. K., Colquhoun, H., Levac, D., Moher, D., Peters, M. D. J., Horsley, T., Weeks, L., Hempel, S., Akl, E. A., Chang, C., McGowan, J., Stewart, L., Hartling, L., Aldcroft, A., Wilson, M. G., Garritty, C., ... Straus, S. E. (2018). Straus, S. E.: Prisma extension for scoping reviews (PRISMA-SCR): Checklist and explanation. *Annals of Internal Medicine*, *169*(7), 467–473. doi:10.7326/M18-0850 PMID:30178033

Williams, M. S., & Patel, V. H. (2022). The impact of covid-19 on the mental health and well-being of immigrant healthcare workers. *Advances in Psychology, Mental Health, and Behavioral Studies*, 34–57. doi:10.4018/978-1-7998-8813-0.ch003

Chapter 4
Resilience and Mental Health Challenges of Climate Refugees:
Adjusting to Forced Relocation in the Current Legal Landscape

Anuttama Ghose
https://orcid.org/0000-0002-7210-4074
School of Law, Dr. Vishwanath Karad MIT-World Peace University, Pune, India

S. M. Aamir Ali
https://orcid.org/0000-0002-8686-0217
Symbiosis Law School Pune, Symbiosis International (Deemed University), Pune, India

ABSTRACT

Climate change-induced catastrophes, which have been shown to have detrimental effects on both physical and mental health outcomes, are anticipated to lead to an increase in human migration beyond existing levels. This chapter examines the complex relationship between resilience and mental health difficulties in the context of climate refugees, who are persons who are forced to relocate as a result of the consequences of climate change. Further, it explores the many aspects of resilience that people experience while facing the psychological consequences of being displaced from their residences. It examines the current legal frameworks that pertain to climate refugees, with a focus on finding any deficiencies and contradictions that contribute to the heightened vulnerabilities experienced by this particular group. This research aims to influence policy conversations and encourage a more holistic approach to addressing the well-being of climate refugees in the face of increasing environmental realities by shedding light on the nexus of resilience, mental health, and legal complexities.

DOI: 10.4018/979-8-3693-2177-5.ch004

INTRODUCTION

In a period characterised by the intensifying complexities of climate change, the growing occurrence of natural calamities, and the relentless ascent of sea levels, individuals displaced due to climate-related factors are emerging as a particularly susceptible and profoundly impacted demographic on a global scale. *Kiribati* and *Tuvalu*, two Pacific states, are now confronted with the pressing issue of sea-level rise, which poses an urgent danger to their territories as the approaching water steadily engulfs their landmasses (Web et al., 2023). The coastal areas of *Bangladesh,* characterized by high population density, are subject to frequent flooding events, resulting in the displacement of a significant number of people (Lázár et al., 2020). The *nation of Syria* had a prolonged period of drought, which, while not directly linked to climate change, exacerbated the shortage of resources and the insecurity of food, so playing a role in the displacement of people inside the country (Selby et al., 2017). Indigenous populations in *Alaska* and *Louisiana* are faced with the challenges of coastal erosion and melting permafrost, which have compelled them to relinquish their historic territories (Maldonado et al., 2021). In the present scenario, island states such as the *Maldives* are confronted with significant issues of an existential kind due to the encroachment of increasing sea levels onto their territorial boundaries (Lelean, 2021). The vulnerability of *Haiti* was intensified as a consequence of the earthquake that occurred in 2010, resulting in further instances of relocation (Llorente-Marrón et al., 2020). Scientists and policymakers have long recognized climate change as a pressing issue that requires urgent attention.

Climate migration is a specialized discipline within the field of migration that pertains to individuals who have been compelled to leave their native lands as a result of the adverse impacts of climate change (Ari & Gokpinar, 2020). These impacts may include rising sea levels and other climate change-induced phenomena such as drought, water scarcity, extreme weather events, deforestation, air pollution, and other climate-related disasters. The phenomenon of migration is expected to undergo significant growth at a rapid rate worldwide in the foreseeable future, unless proactive measures are taken to mitigate the impacts of climate change caused by human activities. Initially, climate migrants were referred to as "environmental refugees" by El-Hinnawi in a United Nations Environment Programme (UNEP) study. El-Hinnawi defines environmental refugees as "*people who have been forced to leave their traditional habitat, temporarily or permanently, because of a marked environmental disruption (natural and/or triggered by people) that jeopardizes their existence and/or seriously affects the quality of their life*" (El-Hinnawi, 1985). This definition encompasses a comprehensive overview of various climate-induced migration patterns and environmental-driven forms of mobility. The question of whether it is preferable to have a definition that is so broad and comprehensive, or if it is challenging to pose a problem with that sort of broad term, is an essential one to explore (Ari & Gokpinar, 2020).

The occurrence of environmental catastrophes compels individuals to leave their residences, occupations, and social networks. Individuals experience both financial losses and intangible psychological impacts (Brammer, 2009). The objective of this study is to investigate the intricate mental health challenges encountered by individuals who have been evacuated due to climate-related factors. This research also investigates the evolving legal variables that impact individuals in question, elucidating the intricate interaction among these several aspects. The phenomenon of migration resulting from climate change has been seen to give rise to significant levels of displacement and deprivation, hence contributing to the experience of stress, anxiety, despair, and trauma among affected individuals. This will also underscore the pressing need to tackle mental health concerns in the context of climate-induced displacement, as well as the remarkable resilience shown by climate refugees in confronting these formidable adversities. Simultaneously, the chapter also tries to analyze the legal determinants that impact the livelihoods of those who are relocated as a consequence of climate change, often known as climate refugees. The recognition is given to the vital importance of legal frameworks in shaping the course of persons' lives, including elements such as protection, entitlements, and freedoms. The objective of this study is to analyze the influence of legal issues on the adaptation and well-being of climate refugees, taking into account the ever-changing nature of the legal landscape. The aim of this research is to ascertain potential strategies for modifying or enhancing existing legislative frameworks to better support individuals in need and adequately address their unique needs. The primary aim of this research is to augment the comprehensive understanding of the complex human dimension of climate change and underscore the need of safeguarding the psychological well-being and legal entitlements of those who are particularly susceptible to its ramifications.

THEORETICAL FOUNDATIONS OF RESILIENCE THEORY IN THE CONTEXT OF CLIMATE REFUGEES AND THEIR MENTAL HEALTH CHALLENGES

The increasing occurrence of displacement caused by climate change has become a significant worldwide concern, requiring a comprehensive comprehension of the mental health difficulties experienced by those who are forced to relocate as a result of environmental deterioration. Resilience theory, which emerged from ecological and psychological viewpoints, offers a comprehensive conceptual framework for understanding how humans negotiate adversity and maintain mental well-being in the face of substantial life disturbances. Resilience theory posits that humans possess intrinsic capabilities to adapt and recover from adversity at its fundamental essence. Originally grounded on ecological principles of system stability, the theory has been expanded to include psychological and social aspects (Southwick et al., 2014). The core principles of this framework include the inherent ability of humans to adapt, the intricate interaction between internal and external stimuli, and the crucial significance of social support networks.

In the context of migration resulting from climate change, persons often referred to as climate refugees encounter significant mental health issues. Resilience theory, which is firmly rooted in the disciplines of psychology and sociology, proves to be an indispensable framework for comprehending and effectively tackling these complex difficulties (Clark-Ginsberg, 2023). The idea explores the intricate dynamics of internal causes, coping strategies, and cognitive evaluation that shape an individual's ability to adapt in the presence of challenging circumstances. Within the context of climate-induced displacement, the significance of individual resilience becomes of utmost importance, necessitating a comprehensive comprehension of how these people effectively manage the psychological consequences of being displaced. Coping mechanisms, such as the active pursuit of social support, the use of problem-solving strategies, and the process of creating meaning from one's experiences, are significant factors that contribute to an individual's level of resilience (Methman & Oels, 2015).

Furthermore, resilience theory expands its scope beyond individual aspects, placing emphasis on the importance of social support networks and community institutions. The preservation of strong social connections, the unity of communities, and the availability of support networks are identified as essential factors that contribute significantly to the resilience of mental health among those displaced by climate-related events (Silchenko & Murray, 2023). The idea further emphasizes the significance of cultural identity and communal bonds in promoting resilience, since the maintenance of cultural traditions and linkages to one's heritage may augment psychological well-being. Resilience theory is noteworthy for its recognition of the possibility of trauma and its introduction of the notion of post-traumatic development. This theory posits that people, given appropriate circumstances and assistance, have the capacity to develop more strength and gain fresh insights in the aftermath of adversity (Clark-Ginsberg, 2023).

In addition, the notion of resilience connects with the concept of environmental justice, hence necessitating an analysis of structural elements that contribute to the relocation of individuals as a result of climate change. The mental health issues of climate refugees might be further intensified by structural inequities associated with environmental deterioration. In summary, resilience theory offers a comprehensive framework that encompasses several aspects such as individual and societal characteristics, coping mechanisms, and the larger systemic environment. This framework informs methods and treatments that attempt to enhance the well-being of those who have been displaced as a result of climate change (Clark-Ginsberg, 2023).

THE LEGAL LACUNA IN THE CONTEXT OF CLIMATE-INDUCED MIGRATION

There is an evident deficiency in protective measures for migrants who are moved across international boundaries due to climate-related catastrophes, with a special emphasis on those originating from island nations and governments with little resources. The existence of this protection gap may be primarily attributed to the absence of a universal entitlement to entrance into a foreign nation under international law. However, those who meet the criteria for protection as refugees or via supplementary measures are safeguarded against being denied access at the border. Consequently, when individuals experience displacement that transcends national boundaries, climate migrants encounter a lack of entitlement to admission into another sovereign nation, the ability to stay inside its borders, or the assurance of safeguarding against involuntary repatriation (Ari & Gokpinar, 2020).

Applicability of International Legal Instruments and 1951 Refugee Convention

The relationship between climate change and human rights is intricate and multifaceted. The most severe repercussions of climate change are now being experienced by the world's least developed countries, which are distinguished by the notable and substantial impacts they are encountering. The insufficiency of preventative measures inside these nations exacerbates these impacts. Nevertheless, it is important to acknowledge that there exists a causal relationship between the effect and the subject under consideration. Populations without sufficient safeguards for the protection of their rights are susceptible to encountering difficulties in understanding or adequately preparing for the consequences of climate change. Individuals may also face challenges while attempting to participate in governmental or multinational endeavors that seek to tackle the issue of climate change. Moreover, it is more probable that these populations have a diminished ability to adjust to projected alterations in their surroundings and ecological systems. The aforementioned scenario engenders an adverse feedback loop that interrelates restricted availability of natural resources, inadequate socioeconomic circumstances, insufficient protections for rights, and increased susceptibility to the negative impacts of climate change.

Many persons and organizations were engaged in the 2007 Bali Conference of the parties, including the United Nations Deputy High Commissioner for Human Rights and personnel of the Office of the High Commissioner for Human Rights (OHCHR). A report on the proceedings was also issued. (UNFCCC, 2008). The effects of climate change on human rights, especially for marginalized groups, are the focus of this investigation. (UNFCCC, 2008). In order to mitigate climate change's negative impacts and safeguard human rights, the study stresses the critical importance of moving swiftly. Equally important is the need to include human rights concerns into plans and policies pertaining to climate change. As part of the continuing global conversation on this pressing topic, the OHCHR has produced an analysis that sheds light on how human rights and climate change interact with one another. (UNFCCC, 2008).

The Human Rights Council passed Resolution 10/4 in March 2009, which addressed the effects of climate change on the exercise of human rights (Human Rights Council, 2009). In order to address the interplay between climate change and human rights, the resolution investigates the possibility of human rights obligations and pledges. Rights to life, adequate nutrition, clean water, and medical treatment are all part of this category of basic human rights. (Human Rights Council, 2009). Furthermore, they could include rights that are threatened by war or forced migration. Because factors like poverty, gender, age, indigenous or minority status, and disability amplify the uneven consequences, the situation's ramifications will hit marginalized groups harder than others.

One of the prominent factors contributing to the insufficient acknowledgment of the distinct social and human rights consequences of climate change, such as migration patterns, is the manner in which it has progressed within the global context. International climate change and human rights governance have evolved separately, grounded on different guiding ideas. Examples of this may be seen in the creation of international environmental treaties that aim to solve these problems, such as the United Nations Framework Convention on Climate Change (UNFCCC) and The International Covenant on Economic, Social and Cultural Rights (ICESCR) (McLeman & Hunter, 2010). When looking at how a human rights treaty is being put into practice, one must take into account how it affects those who are subject to the authority of a state. Everyone knows that when two states interact, especially when it comes to environmental concerns, the government is the one that has to fix any problems that crop up. Countries that are most vulnerable to the negative impacts of climate change are often in a vulnerable position within this domain. However, the involvement of many stakeholders, including governmental and non-governmental entities, tends to complicate the situation and disperse responsibility. The field of human rights law presents challenges in effectively imposing legal responsibilities in situations of this kind. Hence, it is evident that although there is a dearth of a distinct legal response at the global scale, there is an increasingly recognised correlation between climate change and human rights (Ferris, 2017).

Historically, endeavors to acknowledge and assign legal standing to various categories of refugees have mostly been directed within the framework of the refugee convention. But individuals who are compelled to leave their territories due to environmental factors do not satisfy the precise criteria outlined in the 1951 Refugee Convention for the designation of a refugee. The phrase 'climate refugee', which is often used, is therefore considered to be incorrect. Nevertheless, as McAdam points out, while the legal definition of a 'refugee' is specific, it does not imply that those who do not meet this description are undeserving of refugee status or automatically refused it (McAdam, 2012). The 1951 Convention Relating to the Status of Refugees was enacted as an interim solution to alleviate the plight of those displaced by the Second World War. The United Nations (U.N.) had originally designed it with the intention of it being a temporary measure aimed at resolving a transient issue. Nevertheless, in contemporary times, the 1951 Convention and its subsequent 1967 Protocol continue to serve as the primary legal instruments for addressing the plight of refugees worldwide who are compelled to escape their home countries due to persecution (Ghose et. al., 2023). The 1951 Convention, in isolation, is insufficient in addressing the challenges faced by environmentally displaced persons (EDPs) (Anker, 2001). Firstly, its original purpose was to tackle a particular issue, which limits its applicability to the complex circumstances surrounding EDPs. Secondly, the 1951 Convention and 1967 Protocol were formulated without an understanding of climate change, further hindering its ability to adequately meet the needs of EDPs. The moral need to uphold the principle of protection and non-refoulment, which prohibits the forced return of refugees to their countries of origin, should extend to Internally Displaced Persons (IDPs). The causes behind their forced migration vary from those experienced by Convention refugees; nonetheless, the extent of their hardship is comparable (Anker, 2001). Currently, no court has granted international refugee law protection to climate migrants.

Regional Responses to Climate Migration

Certain regional organizations have broadened the scope of the refugee concept to include those displaced due to climate-related factors, hence potentially allowing climate migrants to meet the criteria for refugee status. The Organization of African Unity Convention (OAU) of 1969 provides safeguards for those seeking refuge in a foreign nation due to circumstances such as external attack, occupation, foreign control, or significant disruptions to public order (African Union, 1969). The definition provided in the 1984 Cartagena Declaration encompasses individuals who have been displaced due to various factors, such as the threat to their lives, safety, or freedom resulting from widespread violence, external aggression, internal conflicts, severe human rights violations, or other circumstances that have significantly disrupted public order (UNHCR, 1984). Both of these have been used in instances of environmental migration. The OAU Convention was initiated between the years 2011 and 2012 by the joint efforts of Kenya, Ethiopia, and many East African nations. These countries showed their willingness to accommodate a significant number of Somali refugees, estimated to be over 300,000 individuals, who were compelled to escape their homeland due to the devastating impact of a severe drought (Balk 2019). Ecuador applied the Cartagena Declaration as a framework for accepting Haitian individuals seeking refuge in the aftermath of the seismic event that occurred in 2010 (Carter 2018). However, it is important to note that regional interventions in the context of climate migration are relatively uncommon, making them the exception rather than the rule. As a result, climate migrants often find themselves without adequate legal protection, particularly at the international level.

Fortunately, there has been a growing recognition of the phenomenon of climate-induced migration throughout the international community. The recognition of human mobility within the global climate discourse was initiated with the introduction of the Cancun Adaptation Framework, as part of the United Nations Framework Convention on Climate Change, in the year 2010 (UNFCCC 2010). The Paris Agreement, a significant milestone, played a crucial role in generating this momentum (UNFCCC 2015). As a result, the Task Force on Displacement was established in 2015 under the Warsaw International Mechanism for Loss and Damage associated with Climate Change Impacts. The primary objective of this task force is to prevent, reduce, and manage displacement resulting from the detrimental effects of climate change. The Global Compact on Safe, Orderly, and Regular Migration, also referred to as the Global Compact on Migration, solidified this growing trend by acknowledging climate change as a distinct factor influencing migration in the year 2018 (Global Compact for Migration 2018).

The Global Compact on Migration, which is the first international comprehensive agreement on the protection of migrants, emerged as a result of a collective effort to enhance the availability of legal paths for migration. The Global Compact on Migration addresses the need for enhanced legal safeguards for those who are in transit but do not satisfy the precise criteria outlined in international refugee law to be classified as refugees. The Global Compact on Migration, endorsed by more than 160 countries, proposes various measures to enhance the experience of migration for individuals who do not qualify for protection under international refugee law, such as climate migrants (Global Compact for Migration 2018). These measures include the provision of regular migration pathways, such as humanitarian visas, private sponsorships, and temporary work permits. The aim is to transform migration into a dignified experience rather than a desperate act for those who lack legal protection.

LEGAL INCONSISTENCIES CONTRIBUTING TO VULNERABILITY FOR CLIMATE

Legal inconsistencies that exacerbate vulnerabilities in the context of climate migration are a prominent issue in the contemporary dialogue surrounding climate change. The occurrence of climate-induced migration, which is primarily caused by environmental phenomena such as the escalation of sea levels, severe weather occurrences, and disturbances in agricultural systems, highlights the urgent need for a comprehensive legislative structure. Climate migrants encounter a range of complex difficulties, which arise from the absence of clear international standards and the insufficiency of national laws in effectively addressing the distinct situations associated with forced migration resulting from climate-related consequences.

The Pacific island countries of Kiribati and Tuvalu face significant challenges and legal complexities in relation to climate-induced migration as a result of the imminent danger presented by the rising sea levels. The islands face a significant danger from the expansion of the ocean, which poses a risk of submerging them and rendering them progressively unsuitable for human habitation (Yates et. al., 2023). The impact of environmental changes compels communities to grapple with the need of relocating, resulting in the erosion of deeply rooted cultural connections to the land. Frequently, families with longstanding links to their ancestral residences are driven to relocate from their towns as a result of the immediate hazards presented by climate change.

Notwithstanding these problems, the absence of a legal framework further amplifies the difficulties experienced by the inhabitants of Kiribati and Tuvalu. The absence of a definitive classification for those displaced by environmental conditions within the framework of international law results in a lack of formal recognition for climate refugees (Allgood & McNamara 2017). The lack of a well-defined classification for those displaced by climate change poses challenges in their pursuit of asylum or the acquisition of international protection. As a result, these persons who have been displaced typically do not get the same rights and protections that are granted to conventional refugees, therefore placing them in a vulnerable legal situation. Furthermore, the legal issues transcend the absence of global acknowledgment. The lack of a cohesive international response has led to divergent understandings among various nations on the classification of climate migrants. The classification of these persons by host governments may exhibit inconsistencies, as they may be categorized anywhere along the spectrum from refugees to economic migrants, contingent upon the specific immigration rules of the respective host nations (Allgood & McNamara 2017). The lack of uniformity in legal acknowledgment gives rise to differences in the treatment of those displaced by climate change and obstructs their ability to get essential assistance, such as humanitarian relief and opportunities for resettlement

(Campbell & Warrick 2014). The lack of a unified legal structure intensifies the difficulties encountered by these archipelagic states and their inhabitants, emphasizing the pressing need for all-encompassing and globally acknowledged legal principles to tackle the distinctive conditions of migration caused by climate change.

The Syrian population has been confronted with a lengthy and complex crisis resulting from an extended period of drought. Although the drought cannot be directly ascribed to climate change, it has exacerbated resource shortages and increased the vulnerability of food availability (Eklund et. al., 2022). The protracted period of drought, which occurred between 2006 and 2011, exerted considerable pressure on the nation's water reserves and agricultural capacities. Consequently, rural inhabitants had significant obstacles, such as the deprivation of income, livelihoods, and the capacity to maintain their sustenance. The phenomenon of rural misery served as a spur for internal migration, prompting many people and families to relocate from regions devastated by drought to metropolitan centers in pursuit of improved economic prospects and enhanced living standards (Eklund et. al., 2022). The increased arrival of displaced persons put significant demand on urban resources and exacerbated social and economic strains inside metropolitan areas. Although climate change cannot be directly linked to the occurrence of the drought, it served as a reminder of the vulnerability of places such as Syria to the impacts of climate-related pressures, which have the potential to worsen pre-existing vulnerabilities. Furthermore, the ramifications of the drought were intertwined with wider political and social elements, so adding to the intricate array of events that ultimately precipitated the onset of civil unrest in 2011. The Syrian situation provides a compelling illustration of how environmental stresses, when combined with other socio-political elements, may have extensive ramifications for a country's stability and the welfare of its population. The aforementioned scenario emphasizes the interdependence between climate-related difficulties, limited resources, and societal factors, emphasizing the necessity for comprehensive approaches that tackle both the ecological and socio-political aspects of vulnerability in regions prone to climate-induced pressures (Selby et al., 2017).

The Indigenous populations residing in Alaska, Louisiana, the Maldives, and Haiti exhibit a shared characteristic of increased susceptibility to the impacts of climate change. This vulnerability is further exacerbated by the presence of legal disparities that fail to effectively address the distinct circumstances faced by these communities. The presence of legal loopholes exacerbates environmental, social, and economic issues, rendering these communities more vulnerable to the consequences of migration caused by climate change.

Indigenous groups of Alaska, including the Inuit and Yupik, are now facing the challenges posed by the melting of permafrost, coastal erosion, and the reduction of sea ice as a consequence of increasing temperatures. The alterations have a direct impact on the indigenous communities' customary lifestyles, posing a danger to their traditional hunting and fishing methods (Bronen 2013). The presence of legal contradictions becomes evident when examining the inconsistent protection of the rights of Indigenous peoples to their ancestral lands and resources. Moreover, the absence of well-defined legal frameworks pertaining to displacement caused by climate change poses significant obstacles for Alaskan Indigenous groups in asserting their rights amidst shifting climatic conditions (Milman 2016). Indigenous populations residing in Louisiana, such as the Houma and Biloxi-Chitimacha-Choctaw, are confronted with the challenges posed by the rising sea levels and land erosion, which are further intensified by the impacts of climate change (Jessee 2020). The vulnerability of these communities is exacerbated by legal inconsistencies, since current property laws and catastrophe response procedures fail to fully acknowledge the distinct issues they encounter (Davenport & Robertson 2016).

An ascending sea level threatens the very existence of indigenous Maldivian communities in the archipelago nation of the Maldives. The presence of legal contradictions becomes apparent due to the lack of globally recognized processes pertaining to climate refugees (Kelman et al., 2019). The Maldivian people are confronted with the predicament of displacement caused by climate change, however meet difficulties in finding alternative places of refuge owing to the lack of clear provisions within current legal frameworks pertaining to their position. In Haiti, the Indigenous Taino population, with other marginalized populations, face the challenges posed by severe weather occurrences such as hurricanes and floods, which have been exacerbated by the effects of climate change (Lutz & Yayboke 2021). The insufficiency of disaster response and recovery operations reveals evident legal incongruities, as they often neglect to account for the distinct requirements and entitlements of Indigenous and vulnerable communities. The lack of comprehensive legal frameworks further amplifies the susceptibility of these populations, resulting in their lack of explicit safeguards or means of seeking remedy after climate-induced catastrophes.

The mental well-being of Indigenous people in Alaska, Louisiana, the Maldives, and Haiti is adversely affected by climate change, which is further aggravated by legal disparities. The communities experience considerable psychological suffering as a result of the linked issues they face, including environmental upheaval, loss of traditional livelihoods, and the lack of well-defined legal safeguards. The mental health consequences are exacerbated in every situation due to the convergence of environmental stresses and legal ambiguities. The persistent risk of relocation, erosion of cultural legacy, and insufficiency of legal frameworks in addressing the distinctive difficulties encountered by Indigenous people all lead to a prevailing perception of vulnerability and anguish. To effectively meet the mental health needs of these communities, it is imperative to implement not just environmental and legal solutions, but also culturally responsive mental health support systems that recognise the significant psychological impact of climate change and the resultant legal disparities.

DISRUPTED SOCIAL BONDS, MIGRATION, AND ADVERSE MENTAL HEALTH OUTCOMES

Forced migration entails a range of stressors that individuals experience prior to, during, and following their migration. These stressors encompass various factors such as exposure to violence or catastrophic events, the disruption of familial and social connections, the detachment from one's sense of home, perilous conditions during the journey, risks associated with acculturation, and obstacles in accessing fundamental necessities subsequent to migration. These stresses have the potential to heighten the likelihood of acquiring mental health disorders and may worsen pre-existing diseases. The prevalence of common mental health disorders, including generalized anxiety, major depression, psychosis, suicidality, and Post-Traumatic Stress Disorder (PTSD), as well as other distress symptoms that may not align with Western psychological models, has been found to be higher among forced migrants compared to host populations, according to the World Health Organization (Dunsmore 2022).

The Psychological Ramifications of Climate Disasters

Climate-related calamities, such as floods, hurricanes, and bushfires, have been often linked to mental problems that are influenced by stress. Individuals who have had exposure to potentially life-threatening circumstances are significantly susceptible to the development of posttraumatic stress disorder (PTSD) (DeSalvo et. al., 2007). Post-traumatic stress disorder (PTSD) is characterized by a range of symptoms, which including flashbacks of the traumatic incident, heightened arousal, and deliberate avoidance of stimuli associated with the recollection of the event (McMillen et. al., 2002). In some instances, the symptoms associated with Post-Traumatic Stress Disorder (PTSD) may manifest with a delayed start, often occurring several months to even years subsequent to the individual's exposure to a distressing and even life-threatening event. The occurrence of post-traumatic stress disorder (PTSD) is linked to a decline in the overall well-being and considerable subjective suffering.

Impact of Ambient Temperature on Mental Health

The prevalence of heightened heat exposure is expected to escalate in tandem with the worldwide increase in temperatures. There has been a proposition positing a correlation between an increase in temperature and the manifestation of violent conduct. Moreover, there is a correlation between heat waves and mental and behavioral illnesses (Anderson 2001). According to a research conducted in Australia, there seems to be a correlation between heat waves and heightened rates of hospital admissions for mental disorders, as well as other conditions like cardiovascular and renal illnesses (Nitschke et. al., 2011). Heat waves have been linked to several psychological conditions, including mood disorders, anxiety disorders, dementia, and anxiety-related illnesses, among others. Prolonged exposure to high temperatures may result in both physical and psychological fatigue. According to a research conducted in Thailand, there exists a correlation between occupational heat stress and heightened psychological discomfort among workers (Berry et. al., 2010). Several previous research have also shown a correlation between elevated temperatures in the workplace and heightened levels of psychological discomfort (Epstein & Moran 2006).

Draught Induced Conditions, Economic Transformations, and Escalation of Farmer Suicide

It is quite probable that global climate change will intensify drought occurrences in the foreseeable future. It is anticipated that changes in precipitation patterns may result in heightened occurrences of floods in some regions, while lengthy periods of drought are projected to prevail in other regions (Hanigan et. al., 2012). A correlation has been shown between the incidence of drought and the rate of farmer suicides (Guiney 2012). This phenomenon has been seen not just in wealthy nations such as Australia, but also in emerging countries like India (Deshpande 2002). A correlation has been identified between instances of crop failures resulting from unanticipated drought conditions and incidences of suicide attempts among agricultural workers. The occurrence of crop failure might result in significant economic challenges. In circumstances characterized by limited precipitation, agricultural dependence may render the farmer incapable of maintaining the financial obligations of their household, potentially leading to their entrapment in a cycle of debt as a means to cover these bills. Additionally, this phenomenon might potentially result in an increase in the cost of food and other commodities within the area. The lack of capacity to engage in fundamental transactions may result in malnourishment and susceptibility to further diseases, particularly in poor nations characterized by inadequate organization of structural social services (Sarma 2004). Furthermore, droughts often coincide with extended periods of warm and arid weather conditions. As previously mentioned, there exists a plausible association between heat exposure and heightened frequencies of suicide attempts. Furthermore, extended periods of drought might prompt individuals to relocate to other regions and/or seek alternative occupations. The aforementioned phenomenon gives rise to acculturation stress, which in turn might potentially result in suicide attempts within the agricultural community.

Impact of Disruption of Social Ties

Temporary or long-term separation may be experienced by family and community members when some individuals relocate while others choose to stay in their original locations. Upon being separated, refugees and other individuals who have been displaced may encounter significant administrative obstacles while attempting to reunite with their close family members residing in other countries. In addition to encountering the possible disruption of extended family and community networks, migrants and displaced individuals have difficulties in establishing new social connections within destination areas (L ê 2013). Migrants and displaced individuals often encounter societal stigmatization and discrimination, which therefore leads to their social isolation. The phenomenon of social marginalization has the potential to exacerbate the existing dearth of legal safeguards provided to those who are moved as a result of climatic causes. Consequently, they may encounter many challenges, including heightened levels of stress and the potential development of mental health disorders.

FOSTERING RESILIENCE AND IMPLEMENTING EFFECTIVE RESPONSES TO CLIMATE-INDUCED MIGRATION

The primary objective of adaptation and mitigation methods is to enhance an individual's ability to cope with the dynamic environment, while concurrently striving to minimize the occurrence of environmental changes in the future. The ramifications on mental health are inescapable outcomes of climate change, migration, and movement driven by climate-related factors. Therefore, it is crucial to identify the strategies that governments must include in their policies to effectively address the mental health implications and minimize their impact.

Transition From Traditional Security to Green Security

Given the potential influence of climate change on human mental health, it is crucial to implement strategies aimed at mitigating global warming or developing adaptive measures to address the resulting issues. The mitigation of greenhouse gases entails a reduction in the dependence on fossil fuels, the exploration and use of alternative and efficient sources of energy, the limitation of encroachment on green spaces, and the implementation of comparable strategies. There is an emerging worldwide consensus about the need to decrease the per capita carbon footprint in the next decades, while also addressing the disparities between affluent and impoverished nations. Addressing the issue of climate change necessitates the establishment of collaborative efforts both within and between sectors, as well as on an international scale, in order to effectively execute policies aimed at mitigating the release of greenhouse gases. Developing nations such as India have developed and implemented policies aimed at addressing the adverse effects of climate change (Pandve 2009). The *National Action Plan on Climate Change (NAPCC)* outlines the strategic approach of the Indian government in addressing the challenges posed by climate change. The user did not provide any text to rewrite. The NAPCC encompasses eight missions, namely the *National Solar Mission, National Mission for Enhanced Energy Efficiency, National Mission on Sustainable Habitat, National Water Mission, National Mission for Sustaining the Himalayan Ecosystem, Green India Mission, National Mission for Sustainable Agriculture*, and *National Mission on Strategic Knowledge for Climate Change*. Each mission is directed at the mitigation of climate change processes or the reduction of their effect (Pandve 2009).

Implementation of Resilience Programs: Need of the Hour

There are several characteristics that possess the potential to enhance resilience in the face of a weather-related catastrophe. Coping self-efficacy emerges as a significant predictor in determining psychological effects subsequent to a catastrophic event. Individuals who hold the belief that they possess the ability to effectively manage and adapt to any given circumstance tend to have more favorable outcomes. Furthermore, one of the most robust indicators of favorable mental health outcomes after a catastrophic event is the ability to cultivate and sustain hope, which entails maintaining an optimistic outlook towards the future while experiencing profound sorrow and anguish. Along with that keeping vital healthcare services running during and after harsh weather is crucial (McMichael 2023). Weather-related patient care surges and inpatient sheltering are common for hospitals. Emergency rooms, urgent care centers, labs, and imaging must operate. Residents in nursing homes and residential care institutions are medically weak and vulnerable. Research facilities store valuable samples and data. Additionally, health care facilities may play crucial roles in disaster response by providing clean water, food, and shelter to a broader afflicted population. Public health planners may foster community and social networks by promoting collaboration among schools, churches, and community organizations during physical or psychological challenges. Developing climate preparation strategies with community engagement increases their success rate (McMichael 2023).

Strengthening Legal Safeguards and Introducing Initiatives for Inclusive Mental Health Policies

Within the realm of climate-induced displacement, there exists a pressing need to enhance legal protections and implement strategies that promote comprehensive mental health frameworks. The difficulties encountered by those compelled to move as a result of climate change are not only limited to environmental aspects, but also include significant psychological consequences. Enhancing legal protections entails acknowledging the distinct classification of climate refugees within both global and domestic legal systems, guaranteeing that they get appropriate safeguards and assistance in accordance with their susceptibility. It is essential to concurrently implement measures aimed at fostering comprehensive mental health policy in order to effectively mitigate the psychological impact of relocation. It is imperative that these projects include mental health assistance that is culturally sensitive, care that is informed by trauma, and treatments that are community-based and adapted to the unique experiences of those who have been displaced due to climate-related factors (Torres & Casey 2017). By including legal safeguards with comprehensive mental health measures, communities may provide a more complete and supportive framework for individuals who are facing the intricate difficulties associated with migration caused by climate change.

CONCLUSION

In conclusion, this research work has shed light on the complex and diverse processes that define the experiences of those displaced by climate change. These people are facing physical and mental health challenges as a result of being forced to leave their homes and communities due to the negative effects of climate change. The study's results show that people who are considered climate refugees have certain rights and protections within the legal system that deals with climate-induced migration. Still, the existing rules of law fail to take into consideration the emotional and mental aspects of relocation that these people have to deal with. According to the findings, there is a major loophole in the current legal system that fails to adequately address the psychological and emotional needs of climate refugees.

Recognizing the psychological effects of forced migration and adapting to provide for the holistic welfare of climate refugees, this research stresses the critical need for a paradigm shift in our legal frameworks. Justice, fairness, and the preservation of human rights need that mental health concerns be part of the ever-evolving legal framework. There is a pressing need for this change due to both ethical and practical reasons, since climate-related migration is becoming more common. Rising temperatures are having disproportionate impacts on already vulnerable communities, making it all the more urgent to re-evaluate and expand our legal frameworks to include the mental and emotional toll of relocation. This approach has the potential to improve our capacity to aid climate refugees holistically by taking into account their whole humanity and the intricate web of connections between legal, emotional, and psychological factors in light of the changing world. In order to address the myriad of problems that people face when forced to adapt to a changing environment, this essay makes a compelling case for a legal framework that is compassionate and protective.

REFERENCES

African Union. (1969). *Convention Governing Specific Aspects of Refugee Problems in Africa.* Author.

Allgood, L., & McNamara, K. E. (2017). Climate-induced migration: Exploring local perspectives in Kiribati. *Singapore Journal of Tropical Geography, 38*(3), 370–385. doi:10.1111/sjtg.12202

Anderson, C. A. (2001). Heat and Violence. *Current Directions in Psychological Science, 10*(1), 33–38. doi:10.1111/1467-8721.00109

Anker, D. (2001). Refugee status and violence against women in the domestic sphere: The non-state actor question. *Georgetown Immigration Law Journal, 15*(3), 391–402.

Ari, T., & Gokpinar, F. B. (2020). Climate-Migration: A Security Analysis within the Context of Green Theory. *Uluslararası İlişkiler. International Relations, 17*(68), 41–53. https://www.jstor.org/stable/26980735

Balk, S. (2019). *Deciding Whether to Migrate and Where to Go.* https://disasterdisplacement.org/wp-content/uploads/2019/04/dc_2019-04.pdf

Berry, H. L., Bowen, K., & Kjellstrom, T. (2010). Climate change and mental health: A causal pathways framework. *International Journal of Public Health, 55*(2), 123–132. doi:10.1007/s00038-009-0112-0 PMID:20033251

Brammer, H. (2009). "Climate Refugees": A Rejoinder. *Economic and Political Weekly, 44*(29), 87–87. https://www.jstor.org/stable/40279292

Bronen, R. (2013). *Climate-Induced Displacement of Alaska Native Communities, January 30, 2013.* Alaskan Immigration Justice Project. https://www.brookings.edu/wp-content/uploads/2016/06/30-climate-alaska-bronen-paper.pdf

Campbell, J., & Warrick, O. (2014). *Climate change and migration issues in Pacific*. United Nations Economic and Social Commission for Asia and the Pacific. https://www.ilo.org/dyn/migpractice/docs/261/Pacific.pdf

Cantor, D. J. (2018). *Cross-border Displacement, Climate Change and Disasters: Latin America and the Caribbean*. Platform on Disaster Displacement, UNHCR. https://www.unhcr.org/sites/default/files/legacy-pdf/5d4a7b737.pdf

Clark-Ginsberg, A., & Chandra, A. (2023). Climate change-related mass migration requires health system resilience. *Environmental Research, Health : ERH, 1*(4), 045004. doi:10.1088/2752-5309/ace5ca

Davenport, C., & Robertson, C. (2016, May 2). Resettling the First American 'Climate Refugees'. *The New York Times*. https://www.nytimes.com/2016/05/03/us/resettling-the-first-american-climate-refugees.html

DeSalvo, K. B., Hyre, A. D., Ompad, D. C., Menke, A., Tynes, L. L., & Muntner, P. (2007). Symptoms of posttraumatic stress disorder in a new orleans workforce following hurricane Katrina. *Journal of Urban Health, 84*(2), 142–152. doi:10.1007/s11524-006-9147-1 PMID:17226081

Deshpande, R. S. (2002). Suicide by Farmers in Karnataka: Agrarian Distress and Possible Alleviatory Steps. *Economic and Political Weekly, 37*(26), 2601–2610. https://www.jstor.org/stable/4412301

Dunsmore, C. (2022). *Mitigating Mental Health Impacts of Climate-Related Migration: Incorporating psychological resilience building into disaster risk reduction policy frameworks*. Academic Press.

Eklund, L., Theisen, O. M., Baumann, M., Forø Tollefsen, A., Kuemmerle, T., & Østergaard Nielsen, J. (2022). Societal drought vulnerability and the Syrian climate-conflict nexus are better explained by agriculture than meteorology. *Communications Earth & Environment, 3*(1), 85. doi:10.1038/s43247-022-00405-w

El-Hinnawi, E. (1985). *Environmental Refugees*. United Nations Environmental Programme.

Epstein, Y., & Moran, D. S. (2006). Thermal comfort and the heat stress indices. *Industrial Health, 44*(3), 388–398. doi:10.2486/indhealth.44.388 PMID:16922182

Ferris, E. (2017). Governance and Climate Change-Induced Mobility: International and Regional Frameworks. In D. Manou, A. Baldwin, D. Cubie, A. Mihr, & T. Thorp (Eds.), *Climate Change, Migration and Human Rights: Law and Policy Perspectives* (pp. 49–66). Routledge. doi:10.4324/9781315622217-2

Ghose, A., Bharadwaj, S., & Aamir Ali, S. M. (2023). Politics of Slaughter: A Critical Review of Constitutional Failure in Myanmar and India During Rohingya Crisis. *Novum Jus, 17*(2), 251–277. doi:10.14718/NovumJus.2023.17.2.10

Global Compact for Migration. (2018). *Global Compact for Safe, Orderly and Regular Migration.* https://refugeesmigrants.un.org/sites/default/files/180713_agreed_outcome_global_compact_for_migration.pdf

Guiney, R. (2012). Farming suicides during the Victorian drought: 2001–2007. *The Australian Journal of Rural Health, 20*(1), 11–15. doi:10.1111/j.1440-1584.2011.01244.x PMID:22250871

Hanigan, I. C., Butler, C. D., Kokic, P. N., & Hutchinson, M. F. (2012). Suicide and drought in new south wales, australia, 1970–2007. *Proceedings of the National Academy of Sciences of the United States of America, 109*(35), 13950–13955. doi:10.1073/pnas.1112965109 PMID:22891347

Human Rights Council. (2009). *Tenth Session. Resolution 10/4. Human rights and climate change.* https://ap.ohchr.org/documents/E/HRC/resolutions/A_HRC_RES_10_4.pdf

Jessee, N. (2020). Community resettlement in louisiana: Learning from histories of horror and hope. In S. Laska (Ed.), *Louisiana's Response to Extreme Weather* (pp. 147–184). Springer International Publishing. doi:10.1007/978-3-030-27205-0_6

Kelman, I., Orlowska, J., Upadhyay, H., Stojanov, R., Webersik, C., Simonelli, A. C., Procházka, D., & Němec, D. (2019). Does climate change influence people's migration decisions in Maldives? *Climatic Change, 153*(1–2), 285–299. doi:10.1007/s10584-019-02376-y

Lázár, A. N., Nicholls, R. J., Hall, J. W., Barbour, E. J., & Haque, A. (2020). Contrasting development trajectories for coastal Bangladesh to the end of century. *Regional Environmental Change, 20*(3), 93. doi:10.1007/s10113-020-01681-y

Lê, F., Tracy, M., Norris, F. H., & Galea, S. (2013). Displacement, county social cohesion, and depression after a large-scale traumatic event. *Social Psychiatry and Psychiatric Epidemiology, 48*(11), 1729–1741. doi:10.1007/s00127-013-0698-7 PMID:23644724

Lelean, J. (2021, November 8). The Maldives is threatened by rising seas – but coastal development is causing even more pressing environmental issues. *Connecting Research.* https://research.reading.ac.uk/research-blog/the-maldives-is-threatened-by-rising-seas-but-coastal-development-is-causing-even-more-pressing-environmental-issues/

Llorente-Marrón, M., Díaz-Fernández, M., Méndez-Rodríguez, P., & González Arias, R. (2020). Social vulnerability, gender and disasters. The case of Haiti in 2010. *Sustainability (Basel), 12*(9), 3574. doi:10.3390/su12093574

Lutz, J., & Yayboke, E. (2021, November 22). *Haitian Migration: Food Insecurity, and a Better Way Forward.* Centre for Strategic & International Studies. https://www.csis.org/analysis/haitian-migration-food-insecurity-fragility-and-better-way-forward

Maldonado, J., Wang, I. F. C., Eningowuk, F., Iaukea, L., Lascurain, A., Lazrus, H., Naquin, C. A., Naquin, J., Nogueras-Vidal, K. M., Peterson, K., Rivera-Collazo, I., Souza, M. K., Stege, M., & Thomas, B. (2021). Addressing the challenges of climate-driven community-led resettlement and site expansion: Knowledge sharing, storytelling, healing, and collaborative coalition building. *Journal of Environmental Studies and Sciences, 11*(3), 294–304. doi:10.1007/s13412-021-00695-0 PMID:34099966

McAdam, J. (2012). *Climate Change, Forced Migration, and International Law.* Oxford University Press. doi:10.1093/acprof:oso/9780199587087.001.0001

McLeman, R. A., & Hunter, L. M. (2010). Migration in the context of vulnerability and adaptation to climate change: Insights from analogues. *Wiley Interdisciplinary Reviews: Climate Change, 1*(3), 450–461. doi:10.1002/wcc.51 PMID:22022342

McMichael, C. (2023). Climatic and environmental change, migration, and health. *Annual Review of Public Health, 44*(1), 171–191. doi:10.1146/annurev-publhealth-071421-045148 PMID:36542773

McMillen, C., North, C., Mosley, M., & Smith, E. (2002). Untangling the psychiatric comorbidity of posttraumatic stress disorder in a sample of flood survivors. *Comprehensive Psychiatry, 43*(6), 478–485. doi:10.1053/comp.2002.34632 PMID:12439837

Methmann, C., & Oels, A. (2015). From 'fearing' to 'empowering' climate refugees: Governing climate-induced migration in the name of resilience. *Security Dialogue, 46*(1), 51–68. doi:10.1177/0967010614552548

Milman, O. (2016, December, 19). Alaska indigenous people see culture slipping away as sea ice vanishes. *The Guardian.* https://www.theguardian.com/environment/2016/dec/19/alaska-sea-ice-vanishing-climate-change-indigenous-people

Nitschke, M., Tucker, G. R., Hansen, A. L., Williams, S., Zhang, Y., & Bi, P. (2011). Impact of two recent extreme heat episodes on morbidity and mortality in Adelaide, South Australia: A case-series analysis. *Environmental Health, 10*(1), 42. doi:10.1186/1476-069X-10-42 PMID:21592410

Pandve, H. T. (2009). India's national action plan on climate change. *Indian Journal of Occupational and Environmental Medicine*, *13*(1), 17. doi:10.4103/0019-5278.50718 PMID:20165607

Sarma, E. A. S. (2004). Is Rural Economy Breaking Down? Farmers' Suicides in Andhra Pradesh. *Economic and Political Weekly*, *39*(28), 3087–3089. https://www.jstor.org/stable/4415247

Selby, J., Dahi, O. S., Fröhlich, C., & Hulme, M. (2017). Climate change and the Syrian civil war revisited. *Political Geography*, *60*, 232–244. doi:10.1016/j.polgeo.2017.05.007

Silchenko, D., & Murray, U. (2023). Migration and climate change – The role of social protection. *Climate Risk Management*, *39*, 100472. doi:10.1016/j.crm.2022.100472

Southwick, S. M., Bonanno, G. A., Masten, A. S., Panter-Brick, C., & Yehuda, R. (2014). Resilience definitions, theory, and challenges: Interdisciplinary perspectives. *European Journal of Psychotraumatology*, *5*(1), 25338. doi:10.3402/ejpt.v5.25338 PMID:25317257

Torres, J. M., & Casey, J. A. (2017). The centrality of social ties to climate migration and mental health. *BMC Public Health*, *17*(1), 600. doi:10.1186/s12889-017-4508-0 PMID:28679398

UNHCR. (1984). *Cartagena Declaration on Refugees, adopted by the Colloquium on the International Protection of Refugees in Central America, Mexico and Panama, Cartagena de Indias, Columbia, 22 November 1984*. https://www.unhcr.org/media/cartagena-declaration-refugees-adopted-colloquium-international-protection-refugees-central

United Nation Framework Convention on Climate Change. (2008). *Report of the Conference of the Parties serving as the meeting of the Parties to the Kyoto Protocol on its third session, held in Bali from 3 to 15 December 2007*. https://unfccc.int/resource/docs/2007/cmp3/eng/09.pdf

United Nations Framework Convention on Climate Change (UNFCCC). (2008). *The Cancun Agreements: Outcome of the work of the Ad Hoc Working Group on Long-term Cooperative Action under the Convention*. https://unfccc.int/resource/docs/2010/cop16/eng/07a01.pdf

United Nations Framework Convention on Climate Change (UNFCCC). (2015). *Report of the Conference of the Parties on its 21st session, held in Paris from 30 November to 13 December 2015*. https://unfccc.int/resource/docs/2015/cop21/eng/10.pdf

U.S. Committee for Refugees and Immigrants (USCRI). (n.d.). https://refugees.org/wp-content/uploads/2022/04/Mitigating-Mental-Health-Impacts-of-Climate-Related-Migration.pdf

Web, A., Yusuke, T., Srilata, K., & Pradeep, K. (2023, July 19). *Notes from Tuvalu: Leading the way in adapting to sea-level rise.* PreventionWeb.https://www.preventionweb.net/news/notes-tuvalu-leading-way-adapting-sea-level-rise

Yates, O. E. T., Groot, S., Manuela, S., & Neef, A. (2023). "There's so much more to that sinking island!" - Restorying migration from Kiribati and Tuvalu to Aotearoa New Zealand. *Journal of Community Psychology, 51*(3), 924–944. doi:10.1002/jcop.22928 PMID:36004412

Chapter 5
Resilience Nexus With Climate Change, Food Security, Mental Health, and Social Stability in a Changing World

Revathi Pasupuleti
PVKK Institute of Technology, India

Eswara Reddy Orekanti
Mohan Babu University, India

ABSTRACT

This chapter explores the intricate links between climate change, food security, mental health, and social stability. Key findings reveal that climate change disrupts food security by affecting availability, access, utilization, and stability, impacting regions and populations differently. Mental health is directly and indirectly influenced, leading to trauma, stress, and emotional consequences. Climate change contributes to social instability, and health risks, including conflict and migration due to resource scarcity. Vulnerable populations face unique challenges, necessitating tailored approaches. Improving research is crucial to understand better the complex relationships between climate change and mental health. A holistic approach integrating policy, public health, and global cooperation is essential to address the mental health and social stability implications of climate change for resilience. While mitigation remains vital, proactive adaptation strategies, inclusive of vulnerable populations, are key to reducing adverse impacts on well-being.

DOI: 10.4018/979-8-3693-2177-5.ch005

INTRODUCTION

Climate change is causing significant disruptions in food security worldwide, affecting different regions and populations unequally. This impact is felt through various direct and indirect pathways, with consequences for the availability, accessibility, utilization, and stability of food systems. In response, climate-related food insecurity can lead to or worsen mental health issues by influencing the biological, psychological, social, and cultural factors (Ben Mohamed et al., 2022) that contribute to mental well-being. It's worth noting that these mental health effects of climate change and food insecurity can vary based on factors such as age, gender, and sociocultural context, and they often intersect with other forms of socioeconomic instability.

This paper delves into the intersection of climate change and mental health, specifically examining their impact on ecosystems and food security (Paquin & Falardeau, 2023). It commences by establishing a clear understanding of food security and then delving into the current and anticipated ramifications of climate change on natural systems and the essential components of food security.

It is imperative to acknowledge that climate change stands as one of the most significant health challenges in the 21st century (Zhong et al., 2018). Researchers increasingly recognize its potential to disrupt key factors influencing psychological well-being and overall health. This recognition emphasizes how closely and critically climate change, mental health, and well-being are related.

The social and emotional consequences of climate change on mental health are often elucidated through two primary causal pathways: direct and indirect. The direct impact encompasses the heightened frequency of occurrence and severity of acute or long-term extreme weather events like floods, cyclones, or temperature swings. These events can lead to immediate trauma due to loss of life, property damage, and the onset of acute psychological conditions, including post-traumatic stress disorder.

Indirectly, climate change can influence physical and mental health within societies by disrupting infrastructure, displacing communities, altering the sense of place, and reshaping social environments (Peng, 2019). These environmental transformations, combined with uncertainty about future risks, can trigger a spectrum of emotional responses, from distress and hopelessness to the emergence of clinical mental health disorders.

Climate change's repercussions on food security and ecosystems extend their impact on mental health through both direct and indirect routes that operate at various levels, from the global scale down to individual experiences (Jones et al., 2000). This intricate network of causation underscores the pressing need to address the nexus between climate change and mental health in our endeavors to safeguard well-being in a dynamic and evolving world.

FOOD SECURITY

The concept of food security is a lens through which we examine the intricate relationship between humans and their food sources. This concept came into focus more than half a century ago during a period of global food crises, and it has since evolved as our understanding of the complex factors affecting how people access and consume food has deepened (Shaw, 2007).

A pivotal moment in the development of the food security concept occurred at the 1996 World Food Summit, where a comprehensive definition was put forward to bring unity to various interpretations (Gladwin et al., 2001): "Food security exists when all individuals, at all times, have both the physical and economic means to obtain an adequate supply of safe and nutritious food that caters to their dietary requirements and personal food preferences, supporting an active and healthy life."

Food security has been deconstructed into distinct pillars, each representing a facet of the intricate relationship between humans and their food. To achieve food security, these pillars must be simultaneously upheld. The Food and Agriculture Organization (FAO) of the United Nations, in its efforts to refine this concept, articulated four key pillars in 2008-09, building upon years of evolving understanding. These four pillars encompass:

- **Availability:** This pillar revolves around food production and trade, ensuring that there is an ample supply of food.
- **Access:** It encompasses how people obtain food, their capacity to afford it, and whether the food they access aligns with their sociocultural norms and safety standards.
- **Utilization:** This pillar delves into how the human body processes and assimilates food. It is influenced by factors such as food quality and safety, individuals' overall health, as well as their skills, knowledge, and cultural preferences in food preparation and consumption.
- **Stability:** Stability pertains to the long-term state of these pillars over time. It involves people's ability to consistently access nutritious and culturally appropriate food, which is influenced by society's resilience in facing unexpected shocks, be they natural disasters or socio-economic crises.

These pillars interconnect at various spatial levels, spanning from individual and household levels to regional and global scales (Butt et al., 2005). They also operate across different time frames, as food security can be compromised either as a chronic issue that persists over time or as a transient problem triggered by sudden shocks.

In essence, the concept of food security provides a comprehensive framework to understand the dynamic and multifaceted relationship between humans and their food sources, addressing the complexities of ensuring everyone's access to safe, nutritious, and culturally appropriate sustenance for an active and healthy life.

CLIMATIC CHANGE AND FOOD SECURITY

Several significant climatic changes are exerting their influence on food security, with some of the key drivers being alterations in atmospheric greenhouse gas concentrations, particularly carbon dioxide (CO_2), rising global temperatures, changes in precipitation patterns, and a rise in the number of severe weather conditions. In addition to these, climate pollutants like black carbon and ozone also play a role in shaping food security outcomes.

The effects of climate change on food security are multifaceted and can be categorized as both direct and indirect. They directly affect the availability of food, and these effects can be further compounded by interactions with other environmental stressors, such as pollution, and non-climatic pressures like political crises (Lebel et al., 2022). Importantly, these impacts are not exclusively negative; they can lead to both positive and negative consequences, creating a complex web of effects on food security. Furthermore, these impacts operate at various scales, ranging from localized disruptions to global-scale consequences.

To unravel this intricate web of impacts, it's valuable to consider them within the framework of the four pillars of food security.

- **Food Availability**: Climate change has a direct impact on food availability, primarily through its influence on food production, including crops, livestock, and wild food sources. Climate change can affect the capacity to grow and harvest food (Peng, 2019). Additionally, they can disrupt the storage and transportation of food, with extreme events like floods posing a particular challenge to transport networks.
- **Food Access**: Changes in climate patterns can affect people's ability to access food. Extreme weather events, for example, may disrupt the distribution of food and the functioning of markets, making it harder for individuals to obtain the food they need.
- **Food Utilization**: Climate change can also have indirect consequences for food utilization. Modified weather patterns can have an impact on food safety and quality, as well as how the body absorbs it. Additionally, climate changes may affect the health and well-being of individuals, thereby influencing their ability to utilize food effectively (Self-Brown et al., 2017).

- **Food Stability**: The stability of food security is intricately tied to climate change. Disruptions caused by climate-related events can lead to food shortages and affect the overall resilience of food systems, influencing how people can consistently access food over time (Gregory et al., 2015).

In summary, climate change exerts a substantial impact on food security by affecting its four pillars. It directly influences food availability through changes in production, storage, and transportation, but it can also indirectly affect food access, utilization, and the overall stability of food systems. These effects can manifest at varying scales and often interact with other environmental and non-environmental factors, adding to the complexity of the situation (Sellers et al., 2019).

CLIMATIC CHANGE AND SOCIAL INSTABILITY

In the past, the field of environmental health primarily focused on the identification of potential hazards, the assessment of exposures, and the evaluation of associated risks. This approach involved examining factors that contributed to differential exposure to various environmental threats, including toxic chemicals, pesticides, and radiation. In cases of armed conflict, environmental health experts delved into the specific risks and exposures related to these conflicts, and such investigations were well-documented in the environmental health literature (Fünfgeld, 2015).

Nonetheless, the subject of environmental health has seen a noticeable change in focus in recent years. This shift has placed greater importance on understanding the broader relationships between underlying environmental factors connected to global environmental changes and the resulting insecurities in critical areas such as food, water, and housing (Cook & Bakker, 2012). The traditional paradigm of environmental health, which primarily focused on immediate risks and exposures, is now grappling with the need to address upstream determinants of health and the intricate, long-term interactions between environmental change and human security.

This evolution in environmental health is reflective of a growing awareness that addressing the complex challenges related to global environmental change necessitates a more comprehensive approach, the detailing is shown in Figure 1. It goes beyond merely identifying and quantifying immediate risks to consider the broader factors that contribute to the vulnerability of communities and populations in the face of environmental shifts (Brown, 2004). This expanded perspective recognizes that environmental health is intricately linked to broader issues of human security and well-being, and as a result, new paradigms are emerging to better address these complex, multifaceted dynamics.

Figure 1. A conceptual framework that interconnects climate and environmental systems, physical geography, socioeconomic systems, governance structures, and human health systems (Bourque & Willox, 2014)

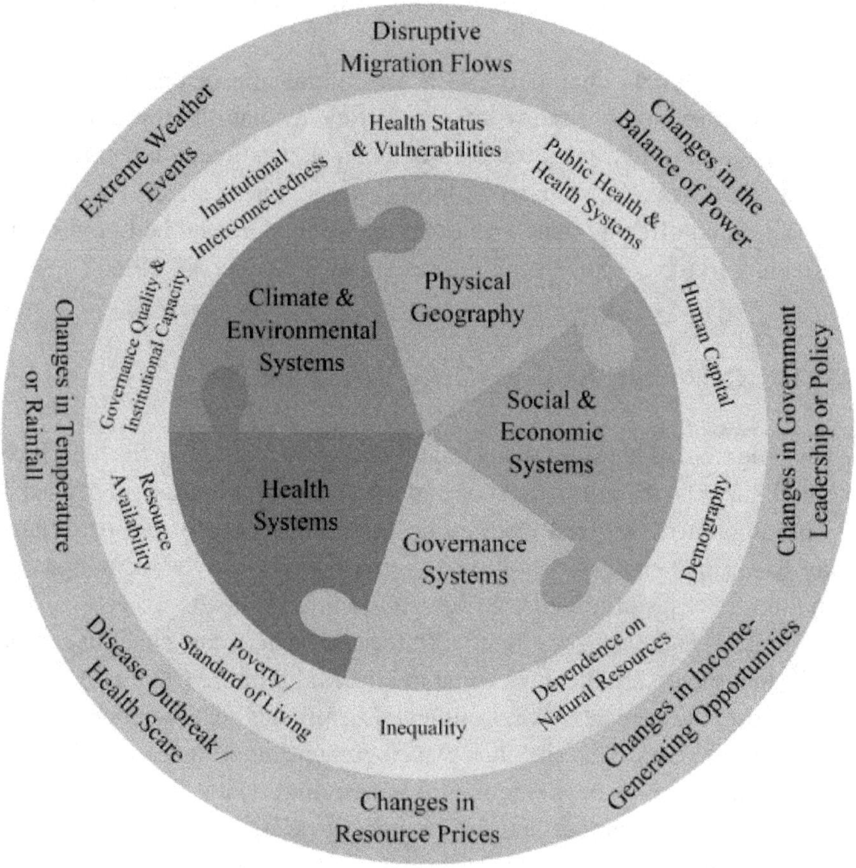

CLIMATIC CHANGE AND MENTAL HEALTH

There is a growing global concern surrounding anthropogenic climate change and the increasing variability in climate patterns. People worldwide are witnessing significant and often dramatic shifts in their environments. Over the 20th century, records have shown a substantial 5°C rise in annual atmospheric temperatures. In the Northern Hemisphere, the rate of increase is particularly striking, with temperatures rising by approximately 1°C per decade.

One stark example of the impact of climate change is the drastic reduction in sea ice extent, as observed in September 2012 when it declined by nearly 50% compared to the 1979-2000 average (Fernandez et al., 2015). This decline marked the lowest coverage on record and served as a powerful indicator of the profound changes occurring in our natural world.

Moreover, there is a growing recognition that these climatic and environmental shifts have the potential to pose challenges to mental health and overall well-being. Research is increasingly shedding light on the emerging global concerns related to mental health issues that might be connected to the alterations and variations in climate (McMichael & Lindgren, 2011). These impacts on mental health can manifest both directly, through the immediate consequences of climate-related events, and indirectly, through the broader ramifications of these changes on societies and individuals, the detailing is shown in Figure 2.

Figure 2. List of the immediate and secondary consequences of climate change and variability on mental health and overall well-being (Füssel, 2009)

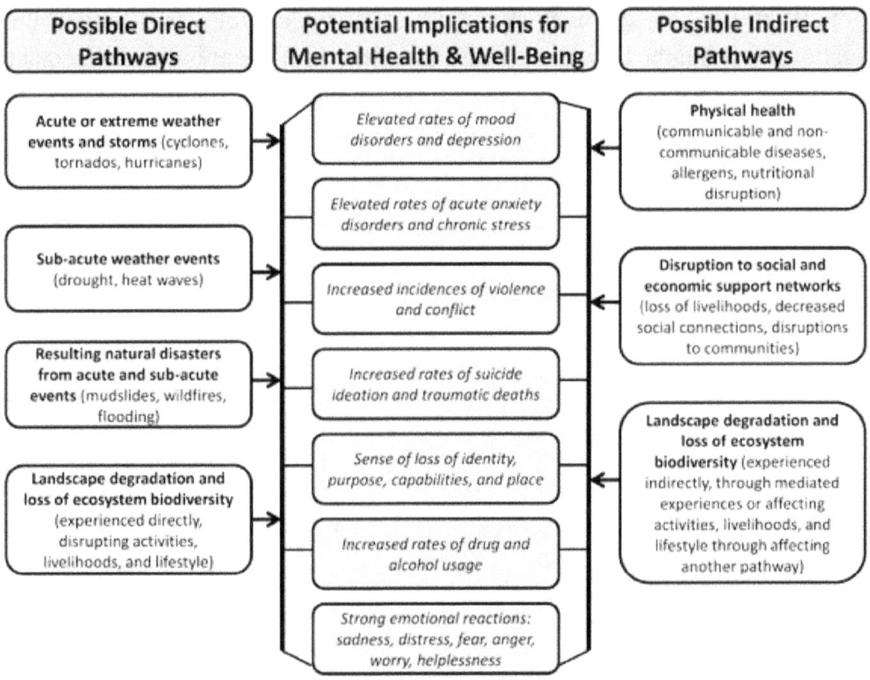

In essence, anthropogenic climate change and its associated variability are transforming the world's climate and environment, and these changes are not only a concern for ecological systems but are also increasingly acknowledged as a factor with profound implications for mental health and overall well-being on a global scale.

The current trajectory of climate change suggests that we are on the brink of witnessing a substantial surge in the number of displaced individuals in the upcoming decades. While we cannot precisely foretell the full scope and magnitude of health risks linked to future climate-induced population movements, analogous situations indicate that these risks are poised to outweigh any potential benefits. This introduces a multifaceted issue with broad geopolitical, ethical, and economic repercussions.

In an ideal scenario, planned resettlement and migration should be regarded as last-resort options. Nevertheless, effective policies should be carefully formulated to streamline the movement of people when it becomes an unavoidable necessity. These policies should be geared towards enhancing well-being and fostering social and economic development both in the regions of origin and destination.

The pivotal question revolves around whether migration can be viewed as an adaptive strategy or if it will be relegated to the status of a last-ditch alternative. This determination hinges on the proactive policy decisions we make today (Bardsley & Hugo, 2010). Beyond self-interest and ethical considerations, there exists a moral obligation rooted in the historical reality that today's affluent industrialized nations bear significant responsibility for the existing buildup of atmospheric greenhouse gases, as shown in Figure 3. Consequently, they should play a substantial role in assisting those affected by climate-related migration.

Figure 3. Environmental health risks: Scale and type – from local direct-acting hazard to global system disruption (Willox et al., 2013)

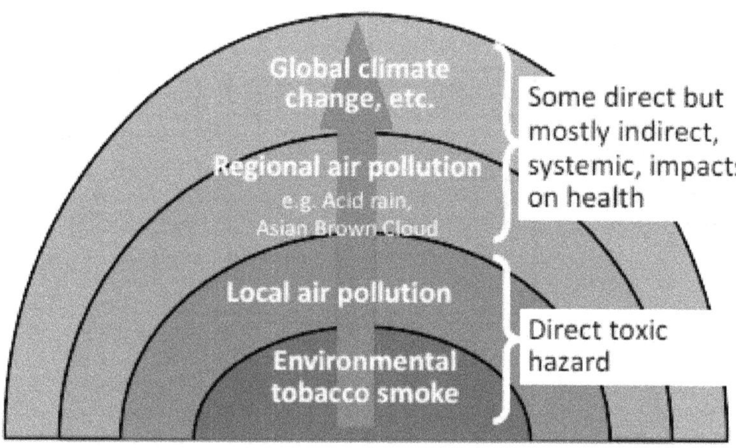

The foremost objective for governments and communities should remain climate change mitigation, which entails the reduction of greenhouse gas concentrations. A comprehensive grasp of the various risks posed by climate change to human well-being and health, including those tied to climate-induced migration, can bolster the international community's resolve to take effective steps for mitigating climate change (McMichael et al., 2012).

Nonetheless, as a certain degree of climate change is already unfolding and future temperature changes are inevitable, adaptive strategies are equally crucial to safeguard human health and well-being. Through a significant reduction in global greenhouse gas emissions and the implementation of comprehensive socio-ecological adaptation measures, the adverse consequences associated with climate-induced migration and its impact on health can be significantly mitigated.

The interplay between climate and conflict is an area of immense policy significance, and it has become a subject of significant attention among researchers worldwide. This review marks the first comprehensive attempt to gather the swiftly expanding body of literature that links climatic shifts to conflict and social instability. What's striking is the remarkable convergence in findings across these studies, especially those that employ robust causal analysis.

These studies, which are best equipped to establish causation, overwhelmingly support the existence of strong connections between climatic anomalies and the emergence of conflict and social instability. Moreover, this growing body of research underscores that climate's impact on security holds both in historical contexts and in our modern era. It's a phenomenon that transcends geographical boundaries and is associated with both rapid and gradual climatic events (Biermann & Boas, 2008). Furthermore, these climatic influences can trigger various forms of conflict at different spatial scales, demonstrating their wide-ranging impact.

The majority of these studies reveal that conflict tends to escalate and social stability tends to decline when temperatures are exceptionally high and precipitation levels are extreme. However, in situations where average temperatures are already temperate, unusually low temperatures can also pose a threat to social stability (Hsiang & Burke, 2014).

The effects of climate change can manifest in diverse ways, whether directly or indirectly, over the short term or long term. Acute climatic events can induce responses akin to traumatic stress, resulting in well-recognized patterns of psychological distress. Moreover, the repercussions of exposure to extreme or prolonged weather-related events can have delayed effects, potentially leading to conditions such as post-traumatic stress disorder. In some cases, these effects may even be passed down to subsequent generations.

In essence, the relationship between climate and conflict is an increasingly well-documented area of research, emphasizing the profound impact of climate on social stability and psychological well-being, with both immediate and far-reaching consequences.

Climate change represents one of the most pressing issues of our time, particularly for the next generation. Its impact on mental health is intricate, with anxiety and depression emerging as prevalent concerns. It's important to recognize that the risk of developing mental health conditions in the aftermath of disasters is not evenly distributed among populations. Extensive research consistently reveals that certain subgroups are more vulnerable than others. Factors such as gender, economic disadvantage, and pre-existing mental health conditions are associated with a heightened susceptibility to mental health issues in the aftermath of disasters (Cianconi et al., 2020). Among these vulnerable groups, children and youth face unique risks due to their close ties to broader social and familial contexts, their ongoing psychological development, and the existential concerns arising from the knowledge that they will contend with the consequences of climate change for many years to come.

To foster better adjustment and coping following disasters, communities should adopt a range of strategies (Hrabok et al., 2020). These may include improving access to mental health care, promoting collaboration among different agencies, enhancing overall resilience, and ensuring adequate preparation for dealing with the mental health impact of climate-related events.

Regarding future research, it is critical to increase the level of methodological rigor in studies evaluating how climate change affects mental health. This can be achieved by incorporating longitudinal research designs, control groups, and more refined assessment methodologies.

The eight Millennium Development Goals were adopted by the United Nations member states in 2000. By 2015, each goal was to have specific, quantifiable targets met, Figure 4 shows the excessive focus on studying climate impact in agriculture over implementing practical solutions. The first goal was to end extreme poverty and hunger, with a particular goal of halving these two issues by the year 2015 in comparison to 1990 levels. This goal underscored the interconnectedness of agriculture, hunger, and poverty, recognizing that poverty limits the ability to cope with these challenges (Von Braun et al., 2005). Improving agricultural productivity is a critical first step in decreasing rural poverty, but in order to achieve rapid rates of growth, other factors like supportive trade and macroeconomic policies, strong infrastructure, and access to credit, land, and markets must also be taken into account. The fight for food security becomes more difficult as a result of climate change, and the poor are disproportionately affected because they have fewer resources to deal with the difficulties it creates.

The results of this study demonstrate that our food systems are being affected by a number of transformative forces, of which climate change is only one. Importantly, the significance of climate change varies not only across different regions but also among diverse societal groups within a given region. It's a complex puzzle where the pieces differ in importance depending on where you look.

Figure 4. A cartoon depicting our exasperation with climate impact studies in agriculture, highlighting minimal advances in understanding while placing far less emphasis on practical solutions and their execution (Campbell et al., 2016)

The paper also underscores that there are options for adapting our food systems to address the challenges posed by climate change. These adaptations can take various forms, involving interventions related to the availability, accessibility, and utilization of food resources (Von Braun et al., 2005). However, it's worth noting that the full assessment of the consequences of these adaptations on the Earth's system is a task that still requires further exploration.

In essence, the paper offers a nuanced perspective on the multifaceted nature of climate change's impact on food systems. It emphasizes the need for region-specific and context-sensitive approaches to adaptation and highlights the ongoing importance of understanding the feedback loops between these adaptations and the broader Earth system.

CONCLUSION

The interconnectedness of climate change, food security, mental health, and social stability is a complex and multifaceted issue that has far-reaching implications for individuals and societies. This topic is of significant global concern, and the review presented highlights several key findings:

1. Climate Change's Impact on Food Security: Climate change disrupts food security by influencing the availability, accessibility, utility, and stability of food systems. All four pillars of food security – availability, access, utilization, and stability – must be tackled concurrently to ensure food security. Climate change, with its direct and indirect effects, can destabilize these pillars, leading to food insecurity that varies across regions and demographics.
2. Climate Change's Influence on Mental Health: Climate change deeply affects mental health through both direct and indirect channels. Direct effects encompass trauma and psychological stress stemming from severe weather conditions and long-term climate shifts. Indirectly, climate change can impact mental well-being by perturbing social and environmental systems, causing displacement, and intensifying uncertainty about the future.
3. Climate Change and Social Disruption: Climate change can contribute to social upheaval, including conflict and migration, due to its effects on resources and living conditions. The importance of proactive policy decisions to address climate-induced migration and displacement is underscored, emphasizing the ethical responsibility of those historically responsible for greenhouse gas emissions.
4. Health Risks Related to Climate Change: The health risks associated with climate change are substantial, covering both direct health impacts and the consequences of climate-related migration. While mitigation remains a primary objective, adaptation strategies are equally necessary to mitigate health risks.
5. Vulnerable Populations: Specific subgroups, such as women, economically disadvantaged individuals, individuals with pre-existing mental health conditions, and children and youth, are particularly susceptible to mental health issues in the context of climate change. Efforts to enhance resilience and coping mechanisms must consider the unique needs of these groups.
6. Future Research Directions: The realm of research on climate change, food security, and mental health is continually evolving. Future research should prioritize enhancements in methodological quality, including longitudinal research designs and control groups, to better comprehend the intricate interplay between climate change and mental health.
7. Global Accountability: Acknowledging the historical contribution of affluent industrialized nations to climate change, there exists a moral obligation for them to assume significant responsibility in aiding those affected by climate-induced migration and displacement.

Mitigating climate change remains the primary goal, but proactive adaptation strategies, inclusive of the most vulnerable populations, are essential to reduce the adverse impacts on human well-being and health.

REFERENCES

Bardsley, D. K., & Hugo, G. J. (2010). Migration and climate change: Examining thresholds of change to guide effective adaptation decision-making. *Population and Environment*, *32*(2-3), 238–262. doi:10.1007/s11111-010-0126-9

Ben Mohamed, A., Van Duivenbooden, N., & Abdoussallam, S. (2002). Impact of climate change on agricultural production in the Sahel–Part 1. Methodological approach and case study for millet in Niger. *Climatic Change*, *54*(3), 327–348. doi:10.1023/A:1016189605188

Biermann, F., & Boas, I. (2008). Protecting climate refugees: The case for a global protocol. *Environment*, *50*(6), 8–17. doi:10.3200/ENVT.50.6.8-17

Bourque, F., & Willox, A. C. (2014). Climate change: The next challenge for public mental health? *International Review of Psychiatry (Abingdon, England)*, *26*(4), 415–422. doi:10.3109/09540261.2014.925851 PMID:25137107

Brown, V. J. (2004). BattleScars: Global conflicts and environmental health. *Environmental Health Perspectives*, *112*(17), A994–A1003. doi:10.1289/ehp.112-a994 PMID:15579406

Butt, T. A., McCarl, B. A., Angerer, J., Dyke, P. T., & Stuth, J. W. (2005). The economic and food security implications of climate change in Mali. *Climatic Change*, *68*(3), 355–378. doi:10.1007/s10584-005-6014-0

Campbell, B. M., Vermeulen, S. J., Aggarwal, P. K., Corner-Dolloff, C., Girvetz, E., Loboguerrero, A. M., & Wollenberg, E. (2016). Reducing risks to food security from climate change. *Global Food Security*, *11*, 34–43. doi:10.1016/j.gfs.2016.06.002

Cianconi, P., Betrò, S., & Janiri, L. (2020). The impact of climate change on mental health: A systematic descriptive review. *Frontiers in Psychiatry*, *11*, 74. doi:10.3389/fpsyt.2020.00074 PMID:32210846

Cook, C., & Bakker, K. (2012). Water security: Debating an emerging paradigm. *Global Environmental Change*, *22*(1), 94–102. doi:10.1016/j.gloenvcha.2011.10.011

Fernandez, A., Black, J., Jones, M., Wilson, L., Salvador-Carulla, L., Astell-Burt, T., & Black, D. (2015). Flooding and mental health: A systematic mapping review. *PLoS One*, *10*(4), e0119929. doi:10.1371/journal.pone.0119929 PMID:25860572

Fünfgeld, H. (2015). Facilitating local climate change adaptation through transnational municipal networks. *Current Opinion in Environmental Sustainability*, *12*, 67–73. doi:10.1016/j.cosust.2014.10.011

Füssel, H. M. (2009). An updated assessment of the risks from climate change based on research published since the IPCC Fourth Assessment Report. *Climatic Change*, *97*(3-4), 469–482. doi:10.1007/s10584-009-9648-5

Gladwin, C. H., Thomson, A. M., Peterson, J. S., & Anderson, A. S. (2001). Addressing food security in Africa via multiple livelihood strategies of women farmers. *Food Policy*, *26*(2), 177–207. doi:10.1016/S0306-9192(00)00045-2

Gregory, P. J., Ingram, J. S., & Brklacich, M. (2005). Climate change and food security. *Philosophical Transactions of the Royal Society of London. Series B, Biological Sciences*, *360*(1463), 2139–2148. doi:10.1098/rstb.2005.1745 PMID:16433099

Hrabok, M., Delorme, A., & Agyapong, V. I. (2020). Threats to mental health and well-being associated with climate change. *Journal of Anxiety Disorders*, *76*, 102295. doi:10.1016/j.janxdis.2020.102295 PMID:32896782

Hsiang, S. M., & Burke, M. (2014). Climate, conflict, and social stability: What does the evidence say? *Climatic Change*, *123*(1), 39–55. doi:10.1007/s10584-013-0868-3

Jones, J. W., Hansen, J. W., Royce, F. S., & Messina, C. D. (2000). Potential benefits of climate forecasting to agriculture. *Agriculture, Ecosystems & Environment*, *82*(1-3), 169–184. doi:10.1016/S0167-8809(00)00225-5

Lebel, L., Paquin, V., Kenny, T. A., Fletcher, C., Nadeau, L., Chachamovich, E., & Lemire, M. (2022). Climate change and Indigenous mental health in the Circumpolar North: A systematic review to inform clinical practice. *Transcultural Psychiatry*, *59*(3), 312–336. doi:10.1177/13634615211066698 PMID:34989262

McMichael, A. J., & Lindgren, E. (2011). Climate change: Present and future risks to health, and necessary responses. *Journal of Internal Medicine*, *270*(5), 401–413. doi:10.1111/j.1365-2796.2011.02415.x PMID:21682780

McMichael, C., Barnett, J., & McMichael, A. J. (2012). An ill wind? Climate change, migration, and health. *Environmental Health Perspectives*, *120*(5), 646–654. doi:10.1289/ehp.1104375 PMID:22266739

PaquinV.FalardeauM. (2023). The complex impacts of climate change on ecosystems, food (in)security, and mental health. doi:10.31234/osf.io/dqjf6

Peng, W. (2019). Construction and application of accounting computerization skills teaching resource database under the background of. *Curriculum and Teaching Methodology*, *2*(1), 1–4. doi:10.23977/curtm.2019.21001

Self-Brown, S., Lai, B., Patterson, A., & Glasheen, T. (2017). The impact of natural disasters on youth: A focus on emerging research beyond internalizing disorders. *Current Psychiatry Reports*, *19*(8), 1–7. doi:10.1007/s11920-017-0798-2 PMID:28681356

Sellers, S., Ebi, K. L., & Hess, J. (2019). Climate change, human health, and social stability: Addressing interlinkages. *Environmental Health Perspectives*, *127*(04), 045002. doi:10.1289/EHP4534 PMID:30986089

Shaw, D. J. (2007). World food summit, 1996. In *World Food Security: A History since 1945* (pp. 347–360). Palgrave Macmillan UK., doi:10.1057/9780230589780_35

Von Braun, J., Swaminathan, M. S., & Rosegrant, M. W. (2005). *Agriculture, food security, nutrition and the Millennium Development Goals. In Annual Report Essays 2004EssayAR*. Intl Food Policy Res Inst.

Willox, A. C., Harper, S. L., Ford, J. D., Edge, V. L., Landman, K., Houle, K., & Wolfrey, C. (2013). Climate change and mental health: An exploratory case study from Rigolet, Nunatsiavut, Canada. *Climatic Change*, *121*(2), 255–270. doi:10.1007/s10584-013-0875-4

Zhong, S., Yang, L., Toloo, S., Wang, Z., Tong, S., Sun, X., & Huang, C. (2018). The long-term physical and psychological health impacts of flooding: A systematic mapping. *The Science of the Total Environment*, *626*, 165–194. doi:10.1016/j.scitotenv.2018.01.041 PMID:29339262

Chapter 6
Selection and Prediction of Sustainable Software Test Bed for Sustainable Test Environment

R. Kamalraj
https://orcid.org/0000-0003-0489-6062
Jain University, India

H. R. Swapna
Jain University, India

Digvijay Pandey
https://orcid.org/0000-0003-0353-174X
Department of Technical Education, IET Lucknow, India

ABSTRACT

The perception of the entire world is to have sustainable products to maintain the green environment. But nowadays the environment temperature is slightly increased in many urban places due to many reasons. One of the reasons is the heavy usage of computer machines for running all commercial applications. To build such applications, the CASE (computer-aided software engineering) tools are used from 'requirements analysis' phase to 'software maintenance' phase. Among all SDLC (software development life cycle) phases, the software testing consumes more power and computing resources for delivering good quality software products to the customer.

DOI: 10.4018/979-8-3693-2177-5.ch006

1. INTRODUCTION

The perception of entire world is to have sustainable products to maintain the Green Environment. But nowadays the environment temperature is slightly increased in many urban places due to many reasons. One of the reasons is the heavy usage of computer machines for running all commercial applications. To build such applications the CASE (Computer Aided Software Engineering) tools are used from 'Requirements Analysis' phase to 'Software Maintenance' phase. Among all SDLC (Software Development Life Cycle) phases, the Software Testing consumes more power and computing resources for delivering good quality software products to the customer. Most of the commercial applications are Web Applications type and the Web Application Testing to be done automatically using Selenium framework. So, a sustainable software testing bed has to be made for automate the testing using automated script with Selenium Framework. A comparative analysis is done on 'PyCharm', 'Python IDLE' and 'Jupyter Notebook' for executing the automated test script using Python Programming platform for identifying the sustainable test bed for web application testing. Along with this, the prediction of test bed performance is also possible by using the datasets made from the past executions and Regression Model. This may help the software engineers and developers to take appropriate decisions with respect to save the globe from 'Global Warming'.

Sustainable software development is highly motivated in the software industry to develop and deliver software systems effectively by consuming less energy withheld the environment free from the heat or carbon emission. By introducing the sustainable software development principles in the software development process thus may help the developers to prepare the plan effectively to fulfill the customer requirements as soon as possible and limit the resources to be used for implementing the project (Winschiers & Paterson,). Software Engineering is a systematic approach to build the software products effectively by applying software development principles, methods and tools. As we know the software development is happening in five different development phases such as 'Requirement Analysis', 'System Design', 'Coding', 'Testing' and finally 'Maintenance' phase. All these phases and their activities are important to deliver a quality software system which can satisfy the expectations of customers and stakeholders.

Nowadays, the software development business is a competitive one because it is happening in small scale industries and/or large scale industries. Many software companies invest huge effort and resources to stand in the software market to dominate others products and gain more customers. Due to the growth of IT industries, huge number of computers and higher end machines are used in worldwide to provide continuous support and services to the users or customers to use the product functions. The higher order machines usage for 24x7 may be a harmful to the environment.

Because it needs continuous cooling system to control the heat in the machine due to use the servers for executing the functions frequently by many number of users. When continuously AC (Air Conditioning) system is working it emits the heat to the environment. Each and every software development companies and schools or colleges who teach the computer science courses have to use of more computers and AC machines in order to maintain the computer machine for a long period of time. But software development cannot be stopped to avoid AC machines usage and huge e-waste from the obsolete computers. The Sustainable goals and objectives are framed from the sustainable development process to use the resources effectively for producing the quality software product (Mahmoud & Ahmad, 2013).

1.1 Sustainable Software Development

Software development is happening based on Software Process Models and Software Development Life Cycle (SDLC) activities. It is an engineering approach where developers apply all principles and policies to build the software product. For that, it may need more resources and efforts. But, the resource consumption may be controlled by connecting with 'Sustainable Software Development Practices' (Johann, 2011).

The following activities are suggested for 'Sustainable Software Development':

- Reducing No. of versions in SRS (Software Requirement Specification) Document. It doesn't mean that to be strict with customers' expectations. It is about plan everything and collects all questions to be used in the Requirements elicitation stage.
- Design error-free Software Designs to propose the solution and help the team how those can be done by software developers.
- Develop a sustainable software code and sustainable software test process or test suites for the problems.
- Propose a sustainable software maintenance process in order to update or modify the system features with sustainable code or elements.

1.2 Sustainable Software Testing

The Software Testing is one of the essential phases in software development to prove the quality of the product and outputs (Aberdour, 2007). To do this, the software testing team should bring testing plan with details about the types of testing would be done on the proposed product and their requirements. Generally, the 'Black-Box', 'White-Box' and 'Integration Testing' will be done in all kinds of project to ensure the functionalities of the software, internal structure and communication among the integrated modules. Hence, testing experts should allocate resources to do these testing types with suitable test suites. But, according to the product type, the testing experts will propose different type of testing types to ensure other unique features of the developed product (Nguyen, 2001). Therefore, the testing team has to find such unique tests for the given product and allocating suitable resources to execute them. Based on number of features or size of the product the testing could be lightweight or heavyweight in terms of number of tests and resources. But mostly the software testing is heavyweight due to conduct of various testing procedures to show the conformity of product features. Software Testing will be done manually and automatically. But in both ways the machines and tools executions are required to collect the results of conducted tests. Due to high usage of tools and machines the Heavyweight Testing may cause heat or more carbon emissions. But the testing procedures cannot be avoided. But it can be normalized by introducing sustainable tools and procedures (Karita et al., 2019).

2. LITERATURE REVIEW

Green Software Development is an essential model to deliver quality software by using the sustainable software tools (Ardito et al., 2015; Mahmoud & Ahmad, 2013). The sustainable software process models are proposed to reduce acquiring the cost and energy during the software development (Agarwal et al., 2012; Calero et al., 2013). The sustainable environment is also one of the major concerns in software industries (Shenoy & Eeratta, 2011). It is completely dependent on selection of sustainable tools from requirements analysis to software maintenance activities. A huge level of resources consumed to produce web applications for providing many commercial applications. Hence, a perfect testing plan should be made for conducting web application testing (Shaukat, 2015). The selection of tools and process are playing vital role to conduct effective testing in web application testing (Raulamo-Jurvanen et al., 2017; Sharma & Angmo, 2014). The 'Selenium' Testing tool is a very popular one in automate the web application testing by having the suitable automated testing script (Gamido & Gamido, 2019; Holmes & Kellogg, 2006). Suggesting a perfect test plan for organizing the 'Sustainable Software Test Process' is required for 'Selenium Testing' with respect to electricity and heat release to the environment.

3. METHODOLOGY FOR SUSTAINABLE TESTING BED

The 'Selenium Testing' is playing a vital role in testing the web applications in terms of their behavior. It could be made as 'Automated' by providing the automated script to trace from the hosting to execution of different features of the web application (Sharma & Angmo, 2014). Nowadays many training or educational institutes are having exercises for 'Selenium Testing' to provide training to students or trainees. Hence, there is a very high usage of the Selenium Testing of the trainees and industry experts. For the mentioned reasons, the selection of test bed or platform for providing training in the Selenium Testing basic is an important one for conducting web application testing. The effect of chosen test bed in environment is to be observed when it is executed by more than 60 no. of students in a training session.

3.1 Environment Setup

The requirements are identified to create the environment for executing the 'Selenium Testing' successfully. The following elements are the requirements for conducting such experiments.

- Programming language for making test script
- Web Browser Driver
- Selenium Package
- Operating Systems
- Tools or IDE
- Web Application

For analysis purpose the python programming platform is chosen for creating automated testing scripts in executing the test cases. Then, the driver software of web browser is required to run the web application through integrating with the automated script. The objects of 'Selenium' package or module should be integrated with the automated script to access the web components developed for the web application. Different tools are available to run the python scripts which executes the 'Selenium' testing on the given web application. The tools considered in this research are 'Python IDLE', 'PYCHARM' and 'Jupyter Notebook'.

The sample python script is given below which is used to open a web application and feed the data automatically to the appropriate data controls. Then the behavior of the application will be monitored by increasing the next level of test cases.

The python script is:

```
from selenium import webdriver
from selenium.webdriver.common.keys import Keys from getpass
import getpass
usr=input("enter username:"); psw=getpass("enter password:");
driver=webdriver.Chrome();
```

```
url = "http://www.localhost:8080/Library" driver.get(url);
username=driver.find_element_by_id("email"); username.send_
keys(usr); password=driver.find_element_by_id("pass");
password.send_keys(psw); password.send_keys(Keys.ENTER);
```

The above python script is written for a web application which is running in a local host to check the behavior of the 'Login' functionality. The URL of a web application running in local host or already hosted application is supplied through a string variable named 'url'. And the 'Chrome' web driver is used to open the application automatically in the 'Chrome' web browser. This above code is executed in 3 different environments made by choosing different IDEs or tools to run python script.

The environments are listed below:

Environment-1: Python IDLE, Selenium Module, Chrome Driver
Environment-2: Jupyter Notebook, Selenium Module, Chrome Driver
Environment-3: Pycharm, Selenium Module, Chrome Driver

The above environments are made in a laptop system which has dual core processor with 16GB of main memory.

3.2 Obtained Results

The three different environments made and those are executed in a month of different days and session. The heat or temperatures of the system processors are observed using a tool 'RealTemp_370'. This tool monitors the execution of processor on doing a particular task. The processors' heat or temperature is collected before and while executing the 'Selenium Testing'. In a session, three different environments are executed and their results are recorded in a worksheet. But after running the Selenium testing in one environment minimum time gap is provided to system processors to come back to their normal stage or before executing Selenium testing. After then, the same script will be executed in environment 2 and 3 to note their performance in terms of processors' heat. The same pattern of observation is done in a month of different 15 days which are in different climate conditions. The maximum processors' temperature in two stages such as 'before testing commencement' and are recorded.

The obtained results when the testing done using 'Python IDLE' is given below.

The above figure shows the temperature of the core processors 'core-1' and 'core-2' as before executing the 'Selenium' testing and testing in 'At Running Stage'. From that, we can get the minimum 6°C temperature difference on core-1 processor and core-2 processor.

Table 1. Python IDLE effect

	Before Running Selenium Testing		At Selenium Testing Running Stage	
Days	**Temperature of Core1**	**Temperature of Core2**	**Temperature of Core1**	**Temperature of Core2**
1	50 °C	52 °C	51 °C	62 °C
2	49 °C	51 °C	59 °C	62 °C
3	47 °C	50 °C	52 °C	57 °C
4	56 °C	60 °C	72 °C	76 °C
5	56 °C	59 °C	59 °C	64 °C
6	50 °C	52 °C	58 °C	61 °C
7	52 °C	56 °C	54 °C	57 °C
8	44 °C	48 °C	52 °C	57 °C
9	44 °C	47 °C	51 °C	57 °C
10	49 °C	53 °C	53 °C	56 °C
11	45 °C	50 °C	55 °C	58 °C
12	44 °C	48 °C	51 °C	55 °C
13	44 °C	48 °C	52 °C	57 °C
14	51 °C	55 °C	57 °C	61 °C
15	57 °C	63 °C	61 °C	64 °C

Figure 1. The average temperature of core processors is found from the above data. The average temperature of core processors before running the testing are 48 and 52, and while running the Selenium testing the maximum temperature of core processors are 54.4 and 58.3.

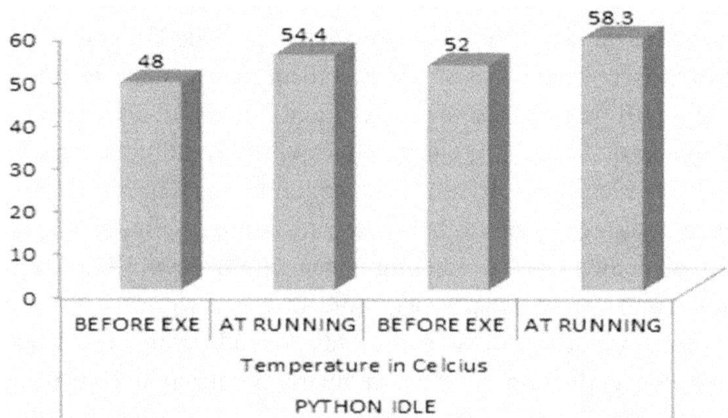

In the same the data are collected for the environment-2 which is done using 'Jupyter Notebook'. The obtained data are shown in the following table-2.

Table 2. Jupyter notebook effect

	Before running Selenium Testing		While Selenium Testing Running	
Days	**Temperature of Core1**	**Temperature of Core2**	**Temperature of Core1**	**Temperature of Core2**
1	56°C	59 °C	67 °C	72 °C
2	49 °C	55 °C	53 °C	58 °C
3	50 °C	55 °C	59 °C	65 °C
4	55 °C	59 °C	58 °C	63 °C
5	54 °C	58 °C	66 °C	70 °C
6	63 °C	66 °C	72 °C	76 °C
7	53 °C	58 °C	59 °C	64 °C
8	56 °C	60 °C	57 °C	63 °C
9	50 °C	55 °C	60 °C	65 °C
10	65 °C	69 °C	68 °C	76 °C
11	60 °C	62 °C	63 °C	66 °C
12	49 °C	53 °C	58 °C	60 °C
13	61 °C	66 °C	61 °C	67 °C
14	60 °C	64 °C	76 °C	76 °C
15	57 °C	58 °C	58 °C	63 °C

Figure 2. The average temperature of core processors before running the testing are 55.9 °C and 59.8 °C, and while running the Selenium testing the maximum temperature of core processors are 62.3 °C and 66.9 °C

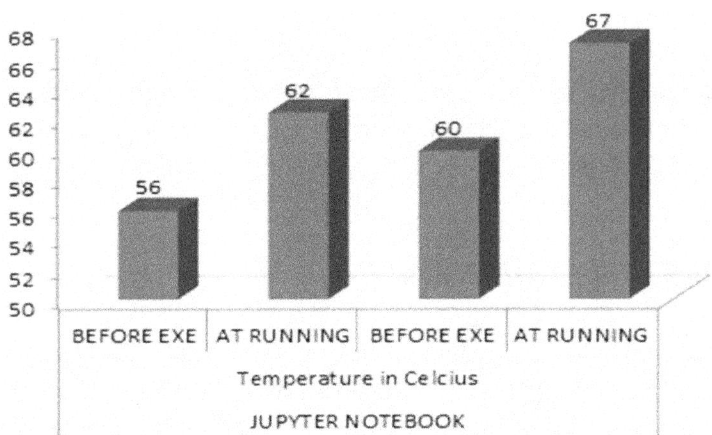

The above bar chart shows that there is 6°C temperature difference on the 'core-1' processor and 7°C difference on the 'core-2' processor.

As like the above, the data collected for the environment-3 which is executing the Selenium Testing using Pycharm tool. The collected temperature data are shown in the following table.

Table 3. PyCharm effect

	Before Running Selenium Testing		While Selenium Testing Running	
Days	**Temperature of Core1**	**Temperature of Core2**	**Temperature of Core1**	**Temperature of Core2**
1	50 °C	54 °C	66 °C	72 °C
2	69 °C	73 °C	52 °C	61 °C
3	34 °C	38 °C	46 °C	47 °C
4	63 °C	68 °C	66 °C	69 °C
5	52 °C	52 °C	54 °C	57 °C
6	59 °C	64 °C	68 °C	73 °C
7	58 °C	63 °C	71 °C	75 °C
8	55 °C	59 °C	58 °C	64 °C
9	53 °C	59 °C	59 °C	65 °C
10	50 °C	53 °C	54 °C	58 °C
11	62 °C	65 °C	67 °C	70 °C
12	57 °C	60 °C	62 °C	66 °C
13	47 °C	51 °C	49 °C	54 °C
14	50 °C	55 °C	58 °C	65 °C
15	53 °C	56 °C	61 °C	65 °C

The average temperature of core processors before running the testing are 54.1 and 58.0, and while running the Selenium testing the maximum temperature of core processors are 59.4 and 64.1.

4. COMPARATIVE ANALYSIS

The collected data from the conducted observations on the 3 different environments for conducting an experiment of 'Selenium Testing' are visualized and analyzed to identify the best option to teach about it. The data are put in the following

Figure 3. Shows that there is around 6°C temperature difference on the 'core-1' processor and 8°C difference on the 'core-2' processor

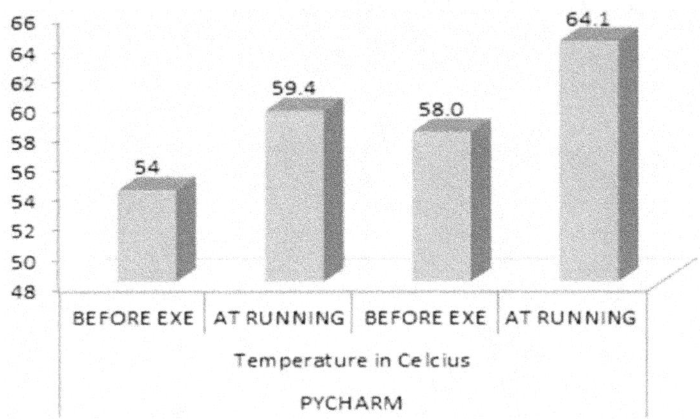

table to do comparative analysis while 'Selenium Testing' is under execution 3 different environments.

To show the difference, the chart is brought from the above tabulated data obtained for different environments.

Table 4. Comparison table for different environments

Python Idle		Pycharm		Jupyter Notebook	
core1	core2	core1	core2	core1	core2
54.4	58.3	59.40	64.1	62	67

In both charts, the 'Trending Lines' are added to show the temperature difference from the environment-1 to environment-3. By analyzing above chart with trending line, we can derive that the 'Environment-1' with 'Python IDLE' produces less heat to the external environment through its 'core- 1' and 'core-2' processors. Then the 'Environment-3' with 'PyCharm', it may produce heat which is greater than 'Envirnoment-1' and lesser than 'Envrionment-2' with 'Jupyter NoteBook' to the external. From this analysis we can recommend 'Environment-1' with 'Python IDLE' based 'Selenium Testing' experiments could be conducted in 'Summer' time and 'Environment-3' with 'PyCharm' or 'Environment-2' with 'Jupyter Notebook' based 'Selenium Testing' experiments could be conducted in the winter season to avoid increase heat in external environment.

Figure 4. The comparisons are done on individual data obtained for each core processor used in the environment. This may help us to come to a conclusion for suggesting the environment for initial level of practices on Selenium Testing.

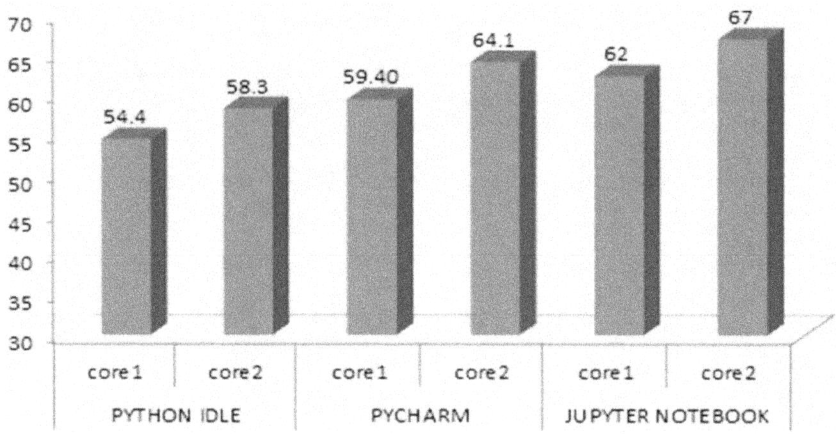

Figure 5. The core-1 processor performance is shown below when it is used on 3 different environments

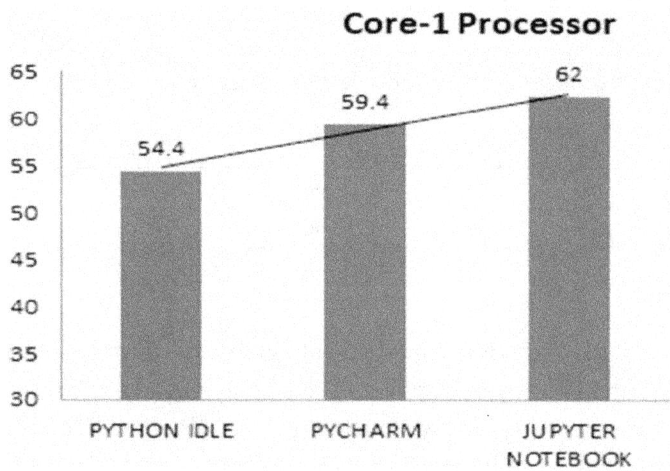

5. REGRESSION ANALYSIS MODEL FOR PREDICTING TOOL SUSTAINABILITY

The 'Selenium Testing' with different platforms is analyzed based on executing a same application with two data controls those will be accessed through automated scripts in different days to get the heat or temperature of the core processors. From

Figure 6. As like that, the core-2 processor temperature difference is shown in the following visualization

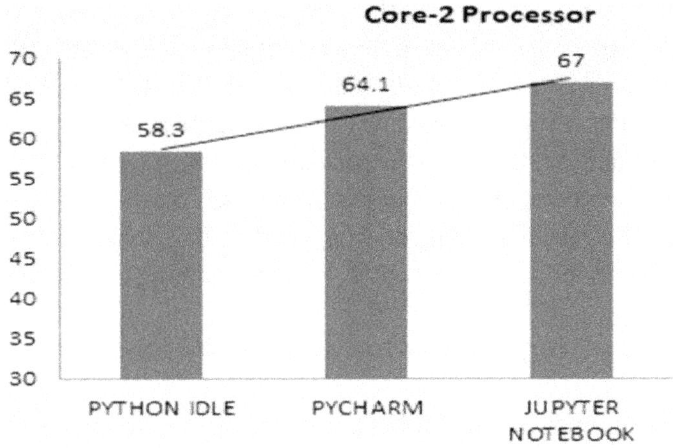

the above analysis we can find the 'Envrionemt-1' with 'Python IDLE' may be producing less heat or temperature and thus help to maintain a green environment. But selection of tool is also based on the performance on testing more than applications with certain no. of data controls interaction in the same script of code. By doing this, there is a possibility of prediction to identify the minimum temperature or heat could be generated by an automated script for testing different web applications (Uyanık & Güler, 2013). To do this, the identified independent parameters are "No. of Applications under Test" and 'No. of Total Data controls processed", and dependent parameter is "Heat or Temperature Produced" from one core processor. A small dataset is created by executing the automated python script with increasing number of applications and data controls to test and obtain the heat or temperature increase of core processor-1.

5.1 Multi Linear Regression for Python IDLE

The obtained dataset for Python IDLE is given in Table-5.

Table 5. Dataset of Python IDLE performance

No. of Applications	X1	1	2	3	4	5	6	7	8
No. of Data controls	X2	2	4	5	7	10	11	13	16
Temperature Increase	Y	7	12	14	14	16	19	21	21

The fields 'X1' and 'X2' are considered to train the 'Multi Linear Regression Model' to predict the temperature when the inputs 'No. of Applications' and 'No. of Data controls' are supplied. By using the model we can predict the temperature of the 'Python Automated Script' which run 9 – No. of applications with 19 – No. of data controls interaction. The model predicts the temperature difference 22.1° C when it is happening in 'Python IDLE'. And if the No. of applications increased to 10 with same 19 – No. of data controls then the predicted temperature is increased to "30.4". From this we can understand that the parameter 'X1" i.e 'No. of Applications" to test having the vital part in deciding the consumption of resources and increasing the 'Temperature" from the current temperature before running the "Selenium Testing".

5.2 Multi Linear Regression for Jupyter Notebook

The collected dataset shown in Table 6 fields are applied to create a 'Multi Linear Regression' model to predict the temperature difference or increase for given 'No. of Applications' (X1) and 'No. of Data Controls' (X2). The created model predicts the temperature difference or increase 20.15° C when X1 is 9 and X2 is 19.

Table 6. The dataset made from 'Jupyter Notebook' by running selenium testing as like done with 'Python IDLE'

No. of Applications	X1	1	2	3	4	5	6	7	8
No. of Data controls	X2	2	4	5	7	10	11	13	16
Temperature Increase	Y	5	14	15	16	18	18	19	19

5.3 Multi Linear Regression for Pycharm

As like Multi linear Regression model on 'Jupyter Notebook' and 'Python IDLE', it is applied for 'PyCharm' IDE platform to identify the temperature increase when various number of applications and data controls interactions happening using python script. The obtained Multi Linear Regression model predicted the temperature increase 19.7° C for given 9 – No. of Applications and 19 – No. of Data controls interactions.

The bar chart in Figure 7 shows the temperature increase of the 'Core Processor-1' when web application testing is executed on different 'Testing Environments'. From the chart we can find that 'PyCharm' environment produced less heat or temperature increase before running and after running the web application tests compared to 'Python IDLE' and 'Jupyter Notebook'. The 'Jupyter Notebook' may be preferred to run more than one web application tests in an automated script because it is also

Table 7. The dataset made from 'Pycharm' by running selenium testing for various applications with their associate data controls interaction using automated python script

No. of Applications	X1	1	2	3	4	5	6	7	8
No. of Data controls	X2	2	4	5	7	10	11	13	16
Temperature Increase	Y	5	13	14	15	16	16	17	19

Figure 7. The difference in temperature increase of the about tested platforms is shown in the below bar chart

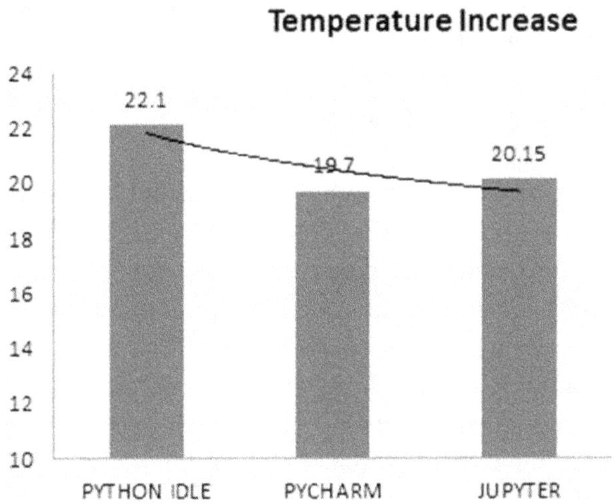

producing almost equal temperature increase as 'Pycharm'. From this we can take a decision that to run more web application tests using 'Selenium' we can prefer 'PyCharm' or 'Jupyter Notebook'. It means that these two environments may be preferred or suggested in 'Software Testing' industries to run more web applications tests in a single automated script.

6. MERITS AND DEMERITS

6.1 Merits

The selection of sustainable tool may help to maintain the environment with less temperature increase whenever huge no. of machines under run the same testing procedure in same time. The conducted study and observation suggests that the

'Python IDLE' may be recommended for running only one web application test using 'Selenium' in technical or training institutes. By doing this the environment temperature or heat may be under control.

A 'Multi Linear Regression' model is made by using the dataset having 2 features such as 'No. of Applications' under test and 'No. of Data Controls' under interaction. This machine learning model used to predict the temperature increase of a processor when number of applications and data controls is increased to conduct web application testing using Selenium. The created model helps to get the predicted temperature increase of web applications test in different environment setups. The conducted prediction suggests the best environment where the temperature increase may be less. This model can be used on industries to choose the best testing environment to maintain a green environment.

6.2 Demerits

The conducted observation is from a single machine but having 2 core processors. The temperature of core processors are collected when testing was done by using 3 different environments on different time periods of a day. The observations have to be done by increasing the number of same configuration of machines.

The 'Multi Linear Regression' model is made with very less size of datasets and dataset is made from only one machine which is used to conduct the test. If data collection done for more than 50 number of same configuration machines then the prediction of the machine learning model may be improved in further level to get accurate prediction.

7. CONCLUSION

The conducted observations on 3 different environment setups based on the tools to run the 'Selenium Testing' help to choose the optimal platform for avoiding the temperature increase in the external environment. There is a possible situation that when around 60 no. of machines runs the same exercises in a particular time unit then there is possibility of increase in environment temperature. But to control that heavy usage of 'Air Conditioner' will be identified. So, selecting a sustainable testing platform may reduce the usage of 'AC' machines and thus may maintain the green environment. The selection of tools is fully based on their sustainability on executing the 'Selenium Testing'. Hence, it may be better to choose the tool which is suitable according to the weather conditions. But, the observation done for sustainability is from a single machine, so this study or observation can be extended to further level as with advanced configurations of machines. And the 'Multi Linear Regression' model for prediction could be improved by adding values of suitable attributes other than 'No. of Applications' and 'No. of Data Controls' of web applications under test. And it is preferred to have a nominal size of dataset which can improve the performance of the proposed model on finding best sustainable testing environment to conduct web application tests on many applications.

REFERENCES

Aberdour, M. (2007). Achieving software quality in open-source software. *IEEE Sustainable Software Testing, 24*(1). 58-64.

Agarwal, Nath, & Chowdhury. (2012). Sustainable approaches and good practices in green software engineering. *International Journal of Research and Reviews in Computer Science, 3*(1).

Ardito, L., Procaccianti, G., Torchiano, M., & Vetrò, A. (2015). Understanding green software development: A conceptual framework. *IT Professional, 17*(1), 44–50. doi:10.1109/MITP.2015.16

Calero, C., Bertoa, M. F., & Moraga, M. Á. (2013). A systematic literature review for software sustainability measures. In *2013 2nd international workshop on green and sustainable software (GREENS)*. IEEE. 10.1109/GREENS.2013.6606421

Gamido & Gamido. (2019). Comparative review of the features of automated software testing tools. *International Journal of Electrical and Computer Engineering, 9*(5).

Holmes, A., & Kellogg, M. (2006). *Automating functional tests using selenium. AGILE 2006 (AGILE'06).*

Johann, T. (2011). Sustainable development, sustainable software, and sustainable software engineering: an integrated approach. In *2011 International Symposium on Humanities, Science and Engineering Research.* IEEE. 10.1109/SHUSER.2011.6008495

Karita, L., Mourão, B. C., & Machado, I. (2019). Software industry awareness on green and sustainable software engineering: a state-of-the-practice survey. *Proceedings of the XXXIII Brazilian Symposium on Software Engineering.* 10.1145/3350768.3350770

Mahmoud, S. S., & Ahmad, I. (2013). A green model for sustainable software engineering. *International Journal of Software Engineering and Its Applications, 7*(4), 55–74.

Nguyen, H. Q. (2001). *Testing applications on the Web: Test planning for Internet-based systems. John Wiley & Sons*. Web Application Testing.

Raulamo-Jurvanen, P., Mäntylä, M., & Garousi, V. (2017). Choosing the right test automation tool: a grey literature review of practitioner sources. *Proceedings of the 21st International Conference on Evaluation and Assessment in Software Engineering*. 10.1145/3084226.3084252

Sharma, M., & Angmo, R. (2014). Web based automation testing and tools. *International Journal of Computer Science and Information Technologies*, *5*(1), 908–912.

Shaukat, K. (2015). Taxonomy of automated software testing tools. *International Journal of Computer Science and Innovation*, *1*, 7–18.

Shenoy, S. S., & Eeratta, R. (2011, December). Green software development model: An approach towards sustainable software development. In *2011 Annual IEEE India Conference* (pp. 1-6). IEEE. 10.1109/INDCON.2011.6139638

Uyanık, G. K., & Güler, N. (2013). A study on multiple linear regression analysis. *Procedia: Social and Behavioral Sciences*, *106*, 234–240. doi:10.1016/j.sbspro.2013.12.027

Winschiers, H., & Paterson, B. (2004). Sustainable software Sustainable Software development. *Proceedings of the 2004 annual research conference of the South Development African institute of computer scientists and information technologists on IT research in developing countries*.

Chapter 7
Smart Sustainability Leveraging Technology to Safeguard Social Stability, Food Security, and Mental Health in a Changing Climate

Rajesh Kanna Rajendran
https://orcid.org/0000-0001-7228-5031
Christ University, Bangalore, India

ABSTRACT

In the face of escalating threats to human health, food security, and ecological sustainability due to climate change, particularly in marginalized and vulnerable communities globally, urgent actions are imperative. This chapter addresses the pressing need for initiatives that mitigate adverse climatic impacts and align with Sustainable Development Goals (SDGs), with a specific focus on food and health outcomes. Recognizing the intricate and multidimensional impact of climate change on these vital aspects of human well-being, there is a growing call for an integrated, science-based approach. The proposed approach advocates for a harmonious marriage of scientific knowledge and local wisdom to address access inequities through agroecology, thereby mitigating the impact of climate change on development-constrained communities. The overarching goal is to pave the way for healthier, more sustainable, and equitable food systems.

DOI: 10.4018/979-8-3693-2177-5.ch007

INTRODUCTION

In an era marked by the profound repercussions of climate change, the interconnected threats to human health, food security, and ecological sustainability have become increasingly urgent. The ramifications of this global phenomenon are particularly pronounced in marginalized and vulnerable communities worldwide, where concerted efforts are imperative to navigate the adverse impacts and align with the Sustainable Development Goals (SDGs), especially in the realms of food and health. The multifaceted challenges posed by climate change demand an integrated, science-based approach that transcends traditional disciplinary boundaries. This paper aims to address this imperative by examining the transdisciplinary practice of agroecology—a holistic and sustainable agricultural paradigm that intricately connects science, on-the-ground practices, and policy. Through a comprehensive analysis of agroecology, this study explores its potential as a pivotal instrument for climate action, particularly in building climate-resilient communities through the establishment of sustainable and equitable food systems. The synthesis of scientific knowledge and local wisdom within the agroecological framework emerges as a promising avenue to mitigate the disproportionate impacts of climate change on vulnerable populations, fostering not only healthier communities but also contributing to the restoration of ecosystems essential for global environmental balance. This introduction sets the stage for an in-depth exploration of the transformative potential of agroecology, advocating for its integration into broader strategic initiatives for climate change adaptation and mitigation within the realms of science and policy.

BACKGROUND STUDY

The global shift toward achieving net-zero emissions necessitates substantial transformations at societal and industrial levels. Governments and corporations are increasingly turning to technological innovations as a key avenue to meet ambitious net-zero targets (Miller, 2020). Within this context, digital technologies emerge as promising tools to address complex societal challenges associated with climate change (George, Merrill, & Schillebeeckx, 2021). A report from the World Economic Forum (WEF), co-authored with PwC and titled "Harnessing Technology for the Global Goals," underscores the significant role that digital technology can play in enhancing resilience to global warming-related hazards, reducing emissions, and empowering humanity to realize net-zero objectives. The report specifically highlights the potential for digital technologies, including Artificial Intelligence (AI), to automate and improve the efficiency of industrial, manufacturing, and agricultural processes, potentially contributing to a 4% reduction in global emissions by 2030 (World Economic Forum & PwC, 2021).

However, the discourse on digital technologies and net-zero goals also acknowledges the negative consequences associated with widespread technological usage, particularly concerning waste products, resource consumption, and CO2 emissions. Instances such as the substantial energy consumption of vast bitcoin mining farms, accounting for 121.36 terawatt-hours per year, underscore the dual nature of rapid technological advancement, presenting both solutions and potential barriers to achieving net-zero targets by 2050 (CBECI, 2021; Mora et al., 2018). Academic literature has started to delve into the debates surrounding the convergence of digital technologies and net-zero imperatives. Studies explore the positive contributions of digital technologies while also adopting a reflective perspective, acknowledging the drawbacks of their rapid adoption (George et al., 2021; Merrill et al., 2019). Despite providing insights into significant challenges and barriers from a digital perspective, the existing discourse remains diverse and disparate, lacking a comprehensive contribution from a wider, more informed, and multi-perspective context. This study aims to fill this gap by offering individual discussions on various interrelated sub-topics within the digital technology and climate change discourse. Each invited expert contributes a unique insight, unraveling the complexities and dependencies inherent in achieving net-zero by 2050.

Social Stability

Climate change-induced events, such as extreme weather conditions and natural disasters, have the potential to disrupt social stability, especially in vulnerable communities. Leveraging technology for early warning systems, community engagement platforms, and disaster response mechanisms becomes imperative. In the Indian context, where densely populated regions are susceptible to climate-related challenges, technology can facilitate efficient communication, evacuation strategies, and resource allocation during crises. Additionally, digital platforms can empower communities to build resilience through knowledge-sharing, creating a more robust social fabric in the face of adversity.

1. **Early Warning Systems:** India is prone to natural disasters such as cyclones and floods. Implementing advanced early warning systems, like the ones used successfully in countries like Japan, can significantly reduce the impact of these disasters. Automated alert systems, accessible through mobile applications or community-specific platforms, can provide timely and accurate information, enabling swift evacuation and minimizing casualties.

2. **Community Engagement Platforms:** Digital platforms can facilitate community engagement and knowledge-sharing. For instance, in the aftermath of a disaster, platforms like Sahana Eden, used globally for disaster management, can be customized to enable communities to coordinate relief efforts, share resources, and communicate effectively.

Food Security

Agriculture, a backbone of the Indian economy, is significantly impacted by climate change, with erratic rainfall patterns, temperature extremes, and changing pest dynamics affecting crop yields. Technology-driven solutions such as precision agriculture, remote sensing, and climate-resilient crop varieties can empower Indian farmers to adapt to these challenges. Mobile applications providing real-time weather forecasts, market information, and farming best practices can enhance decision-making processes. Furthermore, the adoption of blockchain technology in the supply chain can improve transparency, reduce food wastage, and ensure a more secure and efficient food distribution system.

1. **Precision Agriculture:** Climate change affects traditional farming practices in India. Precision agriculture, utilizing technologies like sensors, drones, and AI, can enhance resource efficiency. For example, the use of IoT-enabled soil sensors can provide real-time data on soil conditions, helping farmers make informed decisions about irrigation and fertilization, thereby improving crop yields.
2. **Market Information Systems:** Farmers often face challenges in accessing market information, impacting their bargaining power. Digital platforms such as AgriMarket Mobile App in India provide real-time information on commodity prices, enabling farmers to make informed decisions on when and where to sell their produce.
3. **Blockchain in Supply Chain:** Blockchain technology can be employed to create a transparent and secure food supply chain. This ensures the authenticity of organic produce, reduces wastage, and addresses issues of food fraud. Walmart's pilot project using blockchain for tracking mangoes in India showcases the potential of this technology.

Mental Health

The psychological toll of climate change, amplified by events like cyclones, floods, and heatwaves, necessitates a focus on mental health support. Technology offers avenues for remote mental health services, crisis helplines, and community support networks. In a diverse country like India, where mental health awareness is gradually gaining traction, mobile applications and telemedicine platforms can bridge the gap in accessing mental health resources. Culturally sensitive and language-specific digital mental health interventions can cater to the diverse needs of the population, promoting mental well-being in the face of climate-induced stressors.

1. **Telemedicine and Remote Counseling:** Access to mental health services in rural areas is often limited. Telemedicine platforms like ePsyClinic and iCall, which offer remote counseling, can bridge this gap. These platforms use technology to connect mental health professionals with individuals, providing support in real-time.
2. **Mobile Applications for Stress Management:** Mobile applications tailored to Indian cultural contexts can assist in stress management. For instance, the 'InnerHour' app provides self-help tools and coping mechanisms, making mental health resources accessible to a wider audience.
3. **Community Support Networks:** Online forums and community support networks can be instrumental in creating a sense of belonging and providing a space for individuals to share their experiences. Examples like the "Vandrevala Foundation" in India focus on creating a supportive community for mental health discussions.

Policy Implications

Government policies play a crucial role in fostering the integration of technology into climate resilience strategies. Initiatives like the National Mission for Sustainable Agriculture (NMSA) and the National Health Mission (NHM) in India provide a policy framework for sustainable agriculture and healthcare, but there is a need to further integrate technological solutions.

1. **Incentivizing Innovation:** Policy measures that incentivize businesses and startups working on climate-resilient technologies can accelerate progress. Tax incentives, research grants, and collaboration with academic institutions can stimulate innovation in this space.
2. **Collaborative Governance:** A multi-stakeholder approach involving the government, private sector, and civil society is essential. Platforms like the National Disaster Management Authority (NDMA) need to collaborate with technology companies, researchers, and grassroots organizations to develop effective and localized solutions.

CLIMATE CHANGE: CATALYST FOR COMPLEX CHALLENGES

Climate change is a global phenomenon with far-reaching implications that extend well beyond environmental concerns. It serves as a catalyst for a multitude of complex challenges that affect society on various levels. From disrupting ecosystems and exacerbating extreme weather events to causing resource scarcity and social instability, climate change is a multifaceted issue that demands urgent attention.

Climate Change and Social Stability

The chapter first examines the relationship between climate change and social stability. Increasingly erratic weather patterns, rising sea levels, and resource scarcity intensify competition for essential resources, leading to social unrest, displacement, and conflict. Here, technology plays a pivotal role in early warning systems, conflict resolution, and post-disaster recovery efforts.

Climate Change and Food Security

Next, we explore the intricate connection between climate change and food security. Altered precipitation patterns and temperature extremes disrupt agricultural practices, jeopardizing crop yields and food supplies. Smart agricultural technologies, including precision farming, blockchain-based supply chains, and climate-resilient crop varieties, are revolutionizing the way we produce and distribute food.

Climate Change and Mental Health

This chapter also delves into the often-overlooked mental health implications of climate change. The emotional toll of experiencing climate-related disasters, environmental loss, and existential anxieties cannot be underestimated. Technology offers teletherapy services, mental health apps, and support networks that enable individuals to cope with climate-induced stress and anxiety.

SMART SUSTAINABILITY: A TECHNOLOGICAL APPROACH

In this context, "smart sustainability" emerges as a transformative concept. It signifies the intersection of technology, sustainability, and resilience, offering innovative solutions to address the complex challenges posed by climate change.

Climate-Smart Technologies

Climate-smart technologies, powered by the Internet of Things (IoT), big data analytics, and artificial intelligence (AI), provide real-time climate data and predictive insights. These technologies enable policymakers and communities to make informed decisions about disaster preparedness, resource allocation, and infrastructure development.

Digital Agriculture and Food Security

Digital agriculture solutions are revolutionizing the food supply chain. From soil sensors and weather forecasting models to blockchain-based traceability, these technologies enhance food security by optimizing resource allocation, reducing waste, and ensuring the quality and safety of food products.

Telehealth and Mental Wellness

Telehealth platforms and mental health apps offer accessible and convenient support for individuals facing climate-induced stress and anxiety. AI-driven mental health chatbots provide round-the-clock assistance, while virtual therapy sessions connect individuals with mental health professionals regardless of geographical barriers.

CONCLUSION

The chapter concludes by emphasizing the critical role of technology in achieving smart sustainability. Leveraging climate-smart technologies, digital agriculture, and telehealth services, we can safeguard social stability, food security, and mental health in a changing climate. By embracing these innovations and addressing the accompanying challenges, we have the potential to build a more resilient and sustainable future for all, even in the face of climate change's complex challenges. Smart sustainability represents a beacon of hope, guiding us toward a more resilient and harmonious coexistence with our planet. In navigating the challenges posed by climate change in the Indian context, technology emerges as a powerful ally. By adopting and adapting innovative solutions, India can not only mitigate the impact of climate-related disruptions but also build resilience in social structures, ensure food security, and promote mental well-being. As the nation continues to evolve in its response to a changing climate, the strategic integration of technology becomes imperative for fostering a sustainable and harmonious future.

REFERENCES

George, G., Howard-Grenville, J., Joshi, A., & Tihanyi, L. (2016). Understanding and tackling societal grand challenges through management research. *Academy of Management Journal*, *59*(6), 1880–1895. doi:10.5465/amj.2016.4007

George, G., Merrill, R. K., & Schillebeeckx, S. J. (2021). Digital sustainability and entrepreneurship: How digital innovations are helping tackle climate change and sustainable development. *Entrepreneurship Theory and Practice*, *45*(5), 999–1027. doi:10.1177/1042258719899425

Merrill, Schillebeeckx, & Blakstad. (2019). *Sustainable digital finance in Asia: Creating environmental impact through bank transformation.* SDFA, DBS, UN Environment.

Mikalef, P., Bourab, M., Lekakos, G., & Krogstie, J. (2020). The role of information governance in big data analytics driven innovation. *Information & Management*, *57*(7), 103361. doi:10.1016/j.im.2020.103361

Miller. (2020). *Climate change solutions: The role of technology*. House of Commons Library.

Mora, C., Rollins, R. L., Taladay, K., Kantar, M. B., Chock, M. K., Shimada, M., & Franklin, E. C. (2018). Bitcoin emissions alone could push global warming above 2C. *Nature Climate Change*, *8*(11), 931–933. doi:10.1038/s41558-018-0321-8

Mozaffar, H., & Panteli, N. (2021). The online community knowledge flows: Distance and direction. *European Journal of Information Systems*, 1–14.

Noriega, M. (2020). The application of artificial intelligence in police interrogations: An analysis addressing the proposed effect AI has on racial and gender bias, cooperation, and false confessions. *Futures*, *117*. doi:10.1016/j.futures.2019.102510

Ojo, A. O., & Fauzi, M. A. (2020). Environmental awareness and leadership commitment as determinants of IT professionals engagement in Green IT practices for environmental performance. *Sustainable Production and Consumption*, *24*, 24. doi:10.1016/j.spc.2020.07.017

Pee, L. G., & Pan, S. L. (2022). Climate-Intelligent Cities and Resilient Urbanisation: Challenges and Opportunities for Information Research. *International Journal of Information Management*, *63*, 102446. doi:10.1016/j.ijinfomgt.2021.102446

PittMichaelAbbas. (2021)Public Interest Technology, Citizen Assemblies, and Performative Governance. *IEEE Technology and Society Magazine, 40*(3).

Rowe, F. (2018). Being critical is good, but better with philosophy: From digital transformation and values to future of IS research. *European Journal of Information Systems*, *27*(3), 380–393. doi:10.1080/0960085X.2018.1471789

Chapter 8
Social Media and Its Influence on Students' Intellectual Development in Higher Education

K. V. N. Lakshmi
Jain University, India

N. Meeramani
Jain University, India

H. R. Swapna
Jain University, India

K. Nethravathi
Jain University, India

Asif Hasan
https://orcid.org/0000-0001-9433-3364
Department of Psychology, Aligarh Muslim University, India

Digvijay Pandey
https://orcid.org/0000-0003-0353-174X
Department of Technical Education, IET Lucknow, India

ABSTRACT

Social media technology has become an intrinsic part of the students' intellectual development. The main purpose of using social media is to get the current information. This study aims to provide the social media influence on students intellectual development. This study adopted survey research method to collect the primary data from 468 students. The influence of social media is explained by R-squared, is very low, and with the help of ANOVA test, the significant value was analyzed.

DOI: 10.4018/979-8-3693-2177-5.ch008

1. INTRODUCTION

Social media tools serve as resources to aid in educational processes, facilitate contact between students and faculty, and support knowledge acquisition. It promotes independent and self- driven learning (Vic Lally, Omar Alshehri, 2019). More and more students are using social media platforms to interact, connect, and communicate with others, making social media technology an essential part of their daily life. While social media has the potential to improve education and give kids access to a variety of information, there are also worries about its detrimental effects on mental and academic performance. Understanding how social media technology affects education and students is crucial. Social media has accelerated cultural changes that have positive and harmful outcomes. Information of any kind, whether it is good or bad, may swiftly spread through social media. Students started utilising these social media channels as informational resources and as tools to aid them in finishing their projects, coursework, and other academically connected duties.

Social media is a kind of communication that connects users over the internet. Social media users have the ability to converse, share knowledge, and publish content online. Various social media networks, including Facebook, Instagram, LinkedIn, Twitter, YouTube, Edu Tube, and many more, are accessible for use. Every platform has a unique set of essential components, including a news feed, groups, messenger, photos, videos, business pages, business profiles, images, YouTube Red, and so on. Users can create published and shared video, audio, text, or multimedia compositions using these sites. It gives us the freedom to spend time with our loved ones, learn new things, explore new interests, and have fun. Consequently, social media is used by individuals globally. This research explores the current impact of social media on personality building process of students in Bangalore. This study aims to understand the association between social media and its influence on students' intellectual developments.

2. LITERATURE REVIEW

Reed (2022) looked into how social media use affected executive function. The modern generation is reliant on technology, social media users have addictions, they are attention-seekers, and they use the internet. According to the study, using social media will improve academic communication. Researchers examined how social media impacts the mental health of healthcare professionals (Novera, Connolly, & et al., p. 2022). The future effects of social media on health, especially on mental and physical health as well as relaxation, fueling, and diversion, can be both beneficial and dangerous, he discovered during his research.

Soni (2022) investigated the impact of social media tools for strengthening library and information services. The main objectives are to understand users' perceptions of social media and its tools and to provide services through digital channels. The study found that, social media will have an impact on ICT technologies and be useful for connecting people and facilitating communication. Social media platforms are also influencing our current world to move towards digital development. (Pandi, 2020) conducted a research on impact of social media on postgraduate art students at Alagappa University in Karaikudi, India. The study's goal was to examine department-specific replies and discover user perceptions of social media. According to the survey, utilising social media is more advantageous for users since they can access information from all over the world, make connections with people in other parts of the world, and social media ties are crucial in our modern life.

(Ravinder, 2020) explored social media's management potential, objectives, advantages, and role in the field of digital libraries. The objective was to identify the problems that social media users face and potential solutions, as well as how social media impacts education and learning. The study discovered that because social media is one of the most significant platforms for resource sharing, it will have an impact on knowledge, the knowledge society, education, learning, and research. Egyptian universities' social media tactics and student satisfaction were examined by EASA in 2019. According to this study, social media platforms made users happy but people struggled most with social media's privacy features.

According to Delello's (2015) research, students' use of social media was seen as a useful tool for classroom learning. The results of his study also found students from various academic backgrounds demonstrated increased levels of participation in group projects that involved real- world application for their potential future work.

The discussion above makes clear that several studies on social media use and its effects on students' learning abilities have been conducted. This study is essential since there is an absence of data on how social media affects students' intellectual growth in higher education. The purpose of this study is to determine how social media impacts students' personality development.

3. OBJECTIVES

1. To understand the association between social media and personality building process of students.
2. To study the impact of social media on social awareness and intellectual development of students.

4. SCOPE AND LIMITATION

The present study's objective is to learn more about social media technology and how it affects students' intellectual growth. Only Bangalore-based students of higher education institutions were included in this study. To gather the information needed for this study, the questionnaires were given out to 468 students.

5. HYPOTHESIS

H_{01}: Social media do not impact personality-building process.
H_{02}: Awareness about social media is not associated with gender of respondents.
H_{03}: Social media will not make the students socially aware.
H_{04}: Social media will not make the students Intellectual.

6. METHODOLOGY

In order to accomplish the research objectives for this study, a random sample of students from higher education institutions (N = 468) in Bangalore was employed. The primary source of information is used to gather the data.

6.1 Research Design

Empirical research was conducted to know the influence of social media on students' intellectual development. A well-structured questionnaire was developed to identify the perception of students on social media.

6.2 Sample Size

A total of 468 students are selected on a random basis for data collection.

6.3 Tests Used

- Frequency distribution
- Regression analysis
- ANOVA test

7. RESEARCH RESULTS

Social media can be accessed through many electronic gadgets. The study undertook a study to know the different electronic gadgets used to access the social media. The following frequency table explains the same.

Table 1. Electronic gadgets used for accessing social media

		Frequency	Percent	Valid Percent	Cumulative Percent
Valid	Desktop	4	.9	.9	.9
	Smartphones	400	85.5	85.5	86.3
	Laptop	44	9.4	9.4	95.7
	iPad	12	2.6	2.6	98.3
	Tablets	8	1.7	1.7	100.0
	Total	468	100.0	100.0	

Most of the respondents have opined that smart phones are ideal for accessing social media. Out of total 468 respondents 86% (Apprx.) opined that smartphone is used. Only 4 of them said that Desktop can also be used for accessing social media. Hence it can be concluded that smartphone is the main instrument to access social media.

H_{01}: Social media do not impact personality-building process.

One of the objectives of the study is to analyse whether student's engagement in social media will impact the personality building process. The exposure to different kinds of information may make students to become well equipped with current scenario. To know the impact regression analysis was conducted.

Table 2. Model Summary[b]

Model	R	R Square	Adjusted R Square	Std. Error of the Estimate
1	.729[a]	.532	.531	.848

a. Predictors: (Constant), Social media is more effective
b. Dependent Variable: Personality Building process

The R^2 is 0.532 which is explaining a moderate influence of social media on personality building process. To know the significant value, ANOVA test was conducted.

Table 3. ANOVA[a]

Model		Sum of Squares	df	Mean Square	F	Sig.
1	Regression	381.012	1	381.012	529.903	.000[b]
	Residual	335.065	466	.719		
	Total	716.077	467			

a. Dependent Variable: Personality Building process
b. Predictors: (Constant), Social media is more effective

The significant value being less than 5%, null hypothesis was rejected and concluded that social media impacts personality building process.

H_{02}: Awareness about social media is not associated with gender of respondents.

Table 4. Chi-square tests

	Value	df	Asymp. Sig. (2-sided)	Exact Sig. (2-sided)	Exact Sig. (1-sided)
Pearson Chi-Square	1.478[a]	1	.224		
Continuity Correction[b]	1.066	1	.302		
Likelihood Ratio	1.497	1	.221		
Fisher's Exact Test				.272	.151
Linear-by-Linear Association	1.475	1	.225		
N of Valid Cases	468				

a. 0 cells (0.0%) have expected count less than 5. The minimum expected count is 15.32.
b. Computed only for a 2x2 table

The study was conducted to know whether the awareness level is associated with gender of respondents. The significant value is greater than 5%. Hence the association could not be established and we failed to reject the null hypothesis.

H_{03}: Social media will not make the students socially aware.

Multiple factors would be responsible to make the students socially aware and equip them with societal information. The study was undertaken to know the various factors responsible for this awareness and level of influence placed by them. Social medial influences the respondents in various ways.

Figure 1. To know the extent of influence of these predictor variables on the dependent variable, 'social awareness', regression analysis was conducted

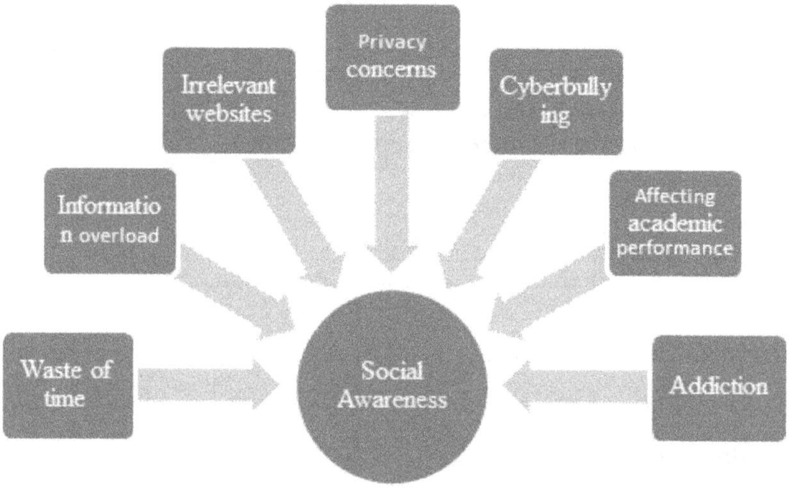

The following table will explain the test results.

Table 5. Model summary[b]

Model	R	R Square	Adjusted R Square	Std. Error of the Estimate
1	.897[a]	.804	.801	.433

a. Predictors: (Constant), Social media creates addiction among student fraternity, social media is the best way to reach information to the students, Social media creates privacy concerns, Information is overloaded through social media, Students perception on social media is a Waste of time, Cyberbullying takes place in social media, Academic performance is affected by social media, Students perception about social media is that Irrelevant websites and ads

b. Dependent Variable: Social awareness of the students

The R^2 is 80%(Apprx.) which is considerable high to explain the influence of predictors on the dependent variable. Hence it can be concluded that there is a considerable influence of these predictor variables in making the students socially aware.

To know the significance of the impact, ANOVA test was conducted.

Table 6. ANOVA[a]

Model		Sum of Squares	df	Mean Square	F	Sig.
1	Regression	354.130	8	44.266	235.706	.000[b]
	Residual	86.201	459	.188		
	Total	440.331	467			

a. Dependent Variable: Social awareness of the students

b. Predictors: (Constant), Social media creates addiction among student fraternity, social media is the best way to reach information to the students, Social media creates privacy concerns, Information is overloaded through social media, Students perception on social media is a Waste of time, Cyberbullying takes place in social media, Academic performance is affected by social media, Students perception about social media is that Irrelevant websites and ads

The ANOVA test is used to know the significance of impact of IV on DV. The high F-value (235.706) and low P-value(<0.05), explains the significant impact. It is concluded that there is a significant impact of the above mentioned predictors on the Dependent variable-social awareness. Hence null hypothesis was rejected.

$\mathbf{H_{04}}$: Social media will not make the students Intellectual.

Table 7. Model summary[b]

Model	R	R Square	Adjusted R Square	Std. Error of the Estimate
1	.145[a]	.021	.004	.591

a. Predictors: (Constant), Social media creates addiction among student fraternity, social media is the best way to reach information to the students, Social media creates privacy concerns, Information is overloaded through social media, Students perception on social media is a Waste of time, Cyberbullying takes place in social media, Academic performance is affected by social media, Students perception about social media is that Irrelevant websites and ads

b. Dependent Variable: Intellectual development of students

It was understood that social media will create societal awareness among students. The study was further conducted to know whether it influences the intellectual development of the students, regression analysis was conducted. The R-squared is 0.021 which explains a very very low influence of IV on the DV. Hence it was concluded that though, social media makes the students socially aware, the extent of addiction and simply browsing for the sake of looking at ads and other forwarded information, it may not develop intellectual growth of students. To know the significance of impact, ANOVA test was conducted.

Table 8. ANOVA[a]

Model		Sum of Squares	df	Mean Square	F	Sig.
1	Regression	3.423	8	.428	1.226	.282[b]
	Residual	160.190	459	.349		
	Total	163.613	467			

a. Dependent Variable: Intellectual development of students

b. Predictors: (Constant), Social media creates addiction among student fraternity, social media is the best way to reach information to the students, Social media creates privacy concerns, Information is overloaded through social media, Students perception on social media is a Waste of time, Cyberbullying takes place in social media, Academic performance is affected by social media, Students perception about social media is that Irrelevant websites and ads

The significant value is 0.282 which is greater than 0.05. Hence, it can be said that these predictors will not influence the intellectual development of students. We failed to reject the null hypothesis.

8. DISCUSSION

The study was conducted to know the different gadgets used to access social media. Frequency analysis was run for the same. From the frequency table it was understood that most of the respondents smart phones to access social media compared to all other media. The respondents are from different demographic background. The hypothesis was developed to know whether social media builds personality of students. Regression analysis was conducted for the same. The R-Squared explained a moderate influence and the significant value being less than five percent, null hypothesis was rejected and concluded that social media influence personality building process of students. To analyze the association between gender and awareness level, chi-square test was conducted. The test results and the significant value explained that there is no much association between the two. Hence it was concluded that there could be other influencers and we failed to reject null hypothesis. There are multiple ways in which social media influence the students. To know the impact of these factors on making the students socially aware, regression analysis was conducted. The various ways in which social media influence the students are: it could be a waste of time, the information may be overloaded which burden the student's understanding level, there could be multiple irrelevant websites which could also support or misguide the students, privacy concerns could also be faced by the users, there could be cyber bullying, this could also negatively influence the academic performance of students and finally it can make students to be more and more addict to it. These were taken and IV and social awareness is taken as DV. With the help of R-squared, the influence was explained. The value was very high enough to say that there is a higher influence of IV on the DV. Hence these associated perceptive attribute of the use of social media has a higher influence on making the students' social aware of various current events. Taking the same as independent variables, it was regressed on how these influence the intellectual development of students, regression analysis was conducted. The influence as explained by R-squared is very low and with the help of ANOVA test, the significant value was analyzed. The higher significant value and lower F-value, explained that there is no significant impact of IV on the DV. Hence it could be concluded that, because of these attributes, the intellectual development of students may not take place. The higher level of addiction and much time spent on browsing the information available in social media may not uplift the intellectual development of students. Hence we failed to reject null hypothesis.

9. CONCLUSION

With the advent of social media and the widespread use of smart phones, it is crucial to determine how social media affects students' intellectual development. Students can exchange information, ideas, and thoughts via social media sites like Facebook, Instagram, and LinkedIn. The majority of today's younger generation is dependent on their mobile devices for social media interaction and communication. Social media, when utilised appropriately, can improve students' overall academic performance, increase their knowledge, and enhance their general welfare by fostering collaboration and communication.

REFERENCES

Al-Rahmi, W. M., & Othman, M. S. (2013). The impact of social media use on academic performance among university students: A pilot study. *Journal of Information Systems Research and Innovation.*

Alam, M. S., & Aktar, H. (2021). The Effect of Social Media on Student Academic Performance: A Case Study at the Islamic University of Bangladesh. *International Journal on Transformations of Media, Journalism & Mass Communication.*

Alnaser, A. S., & Al-Shibly, M. S. (2020). *Impacts of social media and demographical characteristics on university admissions case of Jordanian private universities.* Palarch's Journal of Archeology of Egypt.

Alvi, A. K., Kayani, U. S., & Lakhan, A. B. (2022). *Relationship of social media addiction with academic performance;mediating role of student engagement*. Gomal University Journal of Research. doi:10.51380/gujr-38-02-03

Amin, Z., Mansoor, A., & Hussain, S. R. (2016). Impact of Social Media of Student's Academic Performance. *International Journal of Business and Management Invention* .

Bhavani, K., & Meeramani, N. (2023). Social media technology and its impact on students' learning skills. In R. Suchithra (Ed.), *International conference on Recent Developments in Science and Technology, Engineering and Management*. Academic Press.

Delello, J. A. (2015). Using Social Media as a Tool for Learning: A Multi-Disciplinary Study. *International Journal on E-Learning*, *14*(2), 163–180.

Easa, N. F. (2019). Social Media Strategies and Students' Satisfaction at Egyptian Universities. *International Journal of Customer Relationship Marketing and Management*, *10*(1), 1–16. doi:10.4018/IJCRMM.2019010101

El-Badawy, T. A., & Hashem, Y. (2015). The Impact of Social Media on the Academic Development of School Students. *International Journal of Business Administration*.

Hu, C.-P., Hu, Y., & Yan, W.-W. (2014, October). An empirical study of factors influencing user perception of university digital libraries in China. *Library & Information Science Research*, *36*(3-4), 225–233. doi:10.1016/j.lisr.2013.10.008

Ilahl, R. a. (2019). Digital libraries as learning resources. *Journal of Physics: Conference Series*, *1402*, 1–6.

Kari, H. K. (2019). Promoting library services the social media era exploring use objectives challenges and percieved influence. *Philosophy of Education*.

Khan, R. (2021, October- December). Importance of digital library in education. *International Journal of Research in Library Science*, *7*(4), 102–117. doi:10.26761/IJRLS.7.4.2021.1467

Lubega, D. P. (2019). *Leveraging Social Media in Higher Education: A Case of universities in Uganda*. European Journal of Open, Distance and e-Learning.

M, S. N., & Khan, F. R. (2019). Examining the impact of social media on the academic performances of Saudi students Prince Sattam Bin Abdul University. *Humanities & Social Sciences Reviews*.

Mishra, C. (2019). *Impact of Social Media on Academic Business School Libraries in India: An Empirical Study*. Library Philosophy and Practice.

Namaziandost, E., & Nasri, M. (2019). The Impact of Social Media on EFL Learners' Speaking Skill A Survey Study Involving EFL Teachers and Students. *Journal of Applied Linguistics and Language Research.*

Novera, C. N., & Connolly, R., & P. W. (2006). Past, present and future impact of social media on health workers' mental health: A text mining approach. *Journal of Modelling in Management.*

Pandi, P. P. (2020). *Impact and Usage of Social Media Among the Post graduate students of Alagappa university Karaikudi.* Research gate.

Pawaskar, P., & Penkar, S. a. (2018). Student's Perception on Social Media and Impact on their Education. *International Journal of Creative Research Thought, 6*(1), 341–344.

Rabatseta, B., Maluleka, J. R., & Onyancha, O. B. (2021). Adoption and use of social media in academic libraries in South Africa. *South African Journal of Library and Information Science, 87*(1). Advance online publication. doi:10.7553/87-1-1926

Ravinder, D. (2020). *Social Medias, Skills to Manage, Purposes, Advantages,and their Role in Library Sectors in Digital Era.* IJIRT.

Reed, P. (2022). Impact of social media use on executive function. *Computers in Human Behavior.*

Saaid, S. A., & Wahab, Z. A. (2014). The impact of digital -based materials on Undergraduates' reading habit. *International Journal of Social Science and Humanity, 4*(3), 249–253. doi:10.7763/IJSSH.2014.V4.357

Singh, P. K. (2020). Use of social media platforms with reference to library and information services in digital era. *BIMS International Research of Management and Commerce.*

Soni, S. (2022). Impact of social media tools on enhancing library and information services.

Watson, E. (2020). *Education: The Potential Impact of Social Media and Hashtag Ideology on the Classroom.* Research in Social Sciences and Technology.

Chapter 9
Social Media Crime and Its Impact

Debabrata Singh
Siksha 'O' Anusandhan, India

Swapnarani Mishra
Wipro Pvt. Ltd., India

Anil Kumar Biswal
https://orcid.org/0000-0001-7341-216X
Udayanath Autonomous College of Science and Technology, Cuttack, India

ABSTRACT

In this chapter, the authors analyze online shopping crime, drug trafficking via the dark web, and cyberstalking to illustrate the wide range of problems that arise when trying to criminalize, prevent, and police these activities on social media. An in-depth examination of cybercrime while recharging DTH cases and their newsworthiness in traditional media is a difficult endeavor that is beyond the scope of this research. Instead, this chapter focuses on the process of agenda-building, especially as it relates to social media, the impact of the social environment, and the potential of "ordinary" people to affect agenda definition. They also investigate the reach of the intended audience, the possibility of a "trial by social media," and the function of social media in relation to pre-crime, monitoring, and preventive concerns. Finally, they urge criminologists and social scientists to be more daring and critical in confronting the difficulties provided by new technology.

DOI: 10.4018/979-8-3693-2177-5.ch009

INTRODUCTION

Social media sites have been in the limelight for more than a decade. It's quite useful in regular situations. In addition to their professional impact, the users of these sites have a significant personal one (Milivojevic & McGovern, 2014). This form of social media was originally intended for keeping in touch with friends and family, but it has now evolved into a platform for public debate and conflict resolution on the go.

Social Networking sites are vital communication where individuals can create their public profiles, interact with real-life friends and meet some new people based on their respective interests (Malleson & Andresen, 2015; Ristea et al., 2020). Some popular social networking websites are Facebook, LinkedIn, Instagram, Twitter, Orkut, WhatsApp etc.

Due to their meteoric rise and influence, cybercriminals have begun targeting social media platforms, endangering users' privacy in the process. According to national Investigation Agency, every sixth cybercrime in India is committed through social media. Around 70% rise in cybercrimes annually between 2013 to 2015 according to data provided by NCRB (National Crime Records Bureau) (Müller & Schwarz, 2021; Prieto Curiel et al., 2020). Today cybercrimes are demonstrated in many forms to commit offences related to privacy misrepresentation of identity, cyber terrorism, etc.

BACKGROUND STUDY ON CYBERCRIME

Cybercrime is defined as the criminal activities which takes place on any medium of computers or internet or any other technologies by the Information Technology Act. Cybercriminals are the new attack technologies (Hipp et al., 2019). Online crime is becoming easier through cybercrime-as-a-service and other business schemes. Cybercrime is also called as computer crime; the use of computer for illegal works such as committing fraud, trafficking in child pornography, and intellectual property, stealing identities, or violating privacy.

Cybercriminal activity may be carried out by individuals or groups with relatively little technical skill.

Cybercrime at Odisha

Cybercrime are rapidly increasing as a dominant crime in the society. The highest number of victims who were faced cybercrime were from Odisha (Yang et al., 2018). According National Crime Record Bureau (NCRB), 30 percent of cybercrime cases from Odisha amid the corona virus lockdown. Odisha state registered one of the highest rates of criminal intimidation, online banking scam, ATM fraud, defamation and morphing.

From Fig 1, In 2018 there are 843 cases, in 2019 there are 1475 cases, which can be considered around 76%. In 2020, there are 1931 cases, if we considered between 2020 and 2019 it was 31% more than 2019. In 2021, there are 2065 cases, if we considered between 2021 and 2020 it was 15% more than 2020. In 2022, there are 2450 cases, if we considered between 2022 and 2021 it was 35% more than 2021. In 2023, there are 2995 cases, if we considered between 2023 and 2021 it was 22% more than 2022.

Figure 1. Graphical representation of cybercrime last 3 year records at Odisha

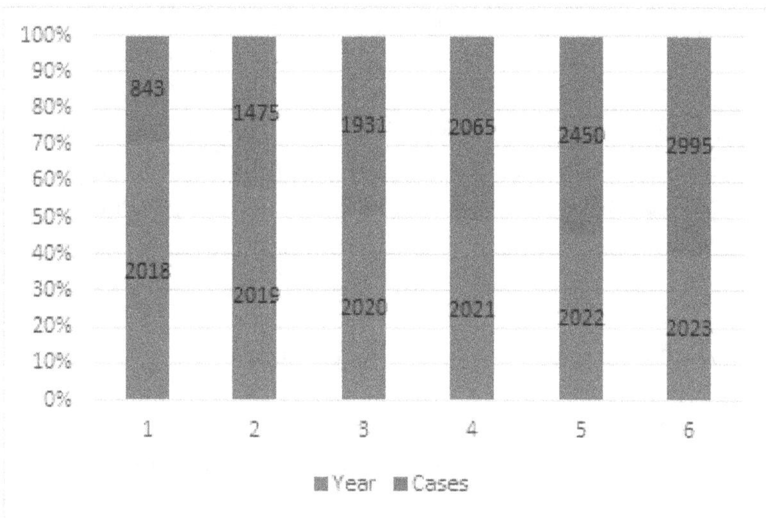

Odisha is in the 4th position in the number of cyber frauds with 1,079 cases, including 132 credit or debit fraud and 549 online banking scams. During corona lockdown period more cases on cyber fraud like online payment, UPI transactions and some are fake video/audio clips related outbreaks the social media. Availability of commodities through different websites made people vulnerable to cyber frauds.

During corona pandemic when all citizen followed "Stay at home", wash hand again and again restrictions of movement of the common people helps the fraudulent to make cyber frauds.

The victims found fraudulent:

1. Withdrawal of money from their accounts by clicked phishing and suspicious mails or links on cash bank offer.
2. Shared their passwords to the cyber crooks.

In last 3 years, more than fourfold increases in number of Cybercrime pending cases & conviction in court. It had been observed that last three years, 220 pending cases in 2019, 548 cases in 2020 and 768 cases in 2021.

The NCRB record indicates the following.

Figure 2. Graphical representation of cybercrime with no. of cases

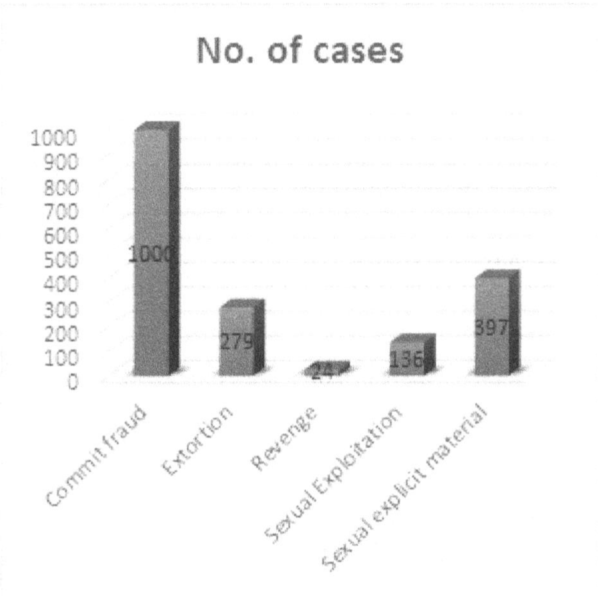

There are many fraud cases increased in the year 2021.

1. ATM fraud: A typical description of ATM fraud is when a criminal makes an unauthorized cash withdrawal using another person's ATM card. Password hacking is used to do this. The third form of ATM fraud I committed was breaking into the ATM itself in order to steal money.
2. Online Banking Fraud: There is a worldwide problem with fraud in online banking (Williams et al., 2020). The term "online banking" refers to financial crimes in which a criminal obtains access to a victim's online bank account

Table 1. Ranks

Rank	State	Cases
1	Bihar	792
2	Maharashtra	454
3	Odisha	331

and then makes unauthorized withdrawals. Internet banking facilitates fraud. A bank is another obvious target for hackers. They can access any account's funds remotely, without ever setting foot inside the bank.

Table 2. Ranks

Rank	State	Cases
1	Maharashtra	552
2	Odisha	545

Main reason of cybercrime:

1. Lack of awareness.
2. Cyber precautions among the victims.

Reasons:

1. Inadequate safety precautions on cyberspace.
2. Strong factor in falling prey to cyber crooks.
3. Mostly women social media user should be careful.
4. Be careful while be friendly with strangers on chatting sites.

India has a total 615.39 cr in more than 1.17 lakh cases of online banking frauds from april 2009 to September 2019.

Cybercrimes at India

There are many cybercrimes increased in India during corona pandemic lockdown (Kilgo et al., 2018). There are 50,035 cases of cybercrime which can be considered around 11.8%. Out of 50,035 cases 578 incidents of fake news in social media.

Table 3. Online banking frauds in 2020

Rank	State	Cases
1	Telengana	1405
2	Maharashtra	821
3	Odisha	549
4	AP	409
5	UP	358
6	Bihar	105

Table 4. Percentage of Internet users in selected countries who have ever experienced any cybercrime of Dec-2019

India	80%
USA	61%
France	60%
New Zealand	59%
Australia	57%
UK	55%
Italy	53%
Netherlands	51%
Germany	47%
Japan	42%

Table 5. The graphical representation of rate of cybercrimes in India

Year	Cases
2018	27248
2019	44735
2020	50035
2021	60450
2022	74554
2023	85462

Figure 3. Graphical representation of cybercrime in year wise cases

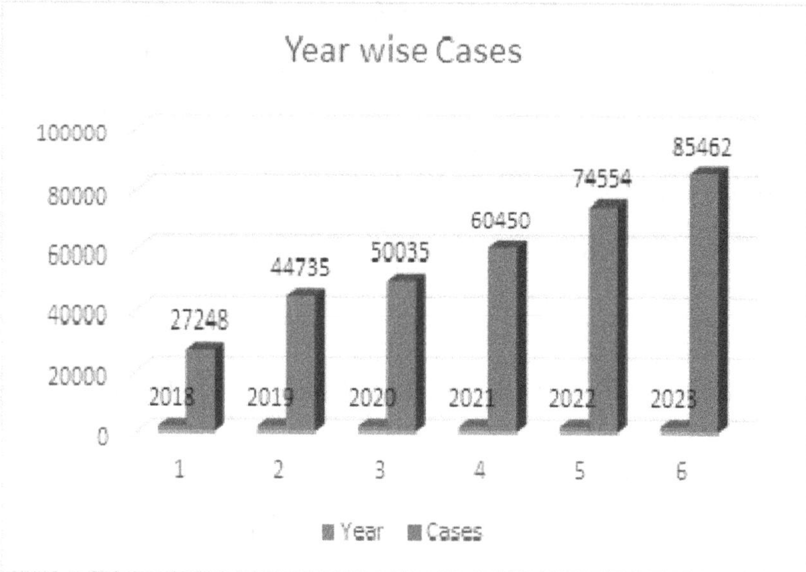

Table 6. In the year 2020, online banking fraud= 4047

Case Name	No. of Cases
OTP fraud	1093
Credit/Debit card fraud	1194
ATM fraud	2160
Fake news	578
Cyber stalking	972
Fake profile	149
Data theft	98

Out of 50,035 case, 30,142(60%) cybercrimes or frauds are lodged. Out of 50,035:

- UP reported 11,097 cases
- Karnataka reported 10,741 cases
- Maharashtra reported 5,496 cases
- Telegana reported 5,024 cases
- Assam reported 3,530 cases

Table 7. Crime rate

State	Rate
Karnataka	16.2%
Telegana	13.4%
Assam	10.1%
UP	4.8%
Maharashtra	4.4%

CASES AND IMPACT OF CYBERCRIME IN SOCIAL MEDIA

Case 1: Software Engineer Encountered Cybercrime While Recharging DTH

Bhubaneswar: Cybercrime is on the rise in the modern world. Appears to be someone who makes a habit of hacking and considers it a way of life (Walsh & O'Connor, 2019). Mostly cybercrime cases are increased from the year 2020. Rising cases of cybercrime in India as well as in World is posing a big challenge.

A girl whose name is Soumya Rout, a software engineer, was allegedly encountered cybercrime of money of Rs 1lakh, while she was recharging her Tv's Direct to Home (DTH) service through an e-commerce payment system named Paytm. According to the reports, she was doing recharge of her Airtel DTH set of box through Paytm wallet on 14th December 2021 in the smart city, Bhubaneswar.

After payment., the message came as Rs 250 was debited from her account and also, she received a cashback of Rs20. But her DTH account was not recharged immediately. Soumya Rout got suspicious and tried to contacting Airtel customer care executive. After contacting Airtel customer care executive, one of the Airtel customer care executive told that they didn't get or receive any recharge of DTH. After talking with Airtel customer care executive, she thought that she should wait more time or maybe within 24 hours to reflect the transaction of her Airtel DTH account.

After waiting for 24 hours, Soumya Rout saw her DTH is still not recharged and she decided to contact Paytm Customer Service to do enquiry about the transaction. Her father, Bairagi Charan Rout, tried to find Paytm Customer Service care number in google and dialed a number. Surprisingly, it was later proved to be a fake number.

Soumya Rout and her father lodged a FIR at the Cyber Police Station. According her FIR report, the following the deduction of money, Soumya Rout called the bank and blocked her account. However, it's too late and more damage already done. She and her father rushed to the police station to file a complaint about her stolen money through Internet.

Soumya Rout somehow finally managed to get the actual Paytm Customer Service number and tried to calling them. They revealed that the entire money is dishonestly taken money off by the recipient(fraudster) and they can reveal the name of the recipient after getting a formal request from the police.

According to media, Bairagi Charan Rout, the father of Soumya Rout said that alleged action are not being taken by the police or by Paytm e-commerce payment company for the recovery of the money stolen by fraudster.

Case 2: Online Shopping Scam: 2 Held for Rs 12-lakh Fraud in Ahmedabad

Ahmedabad: Nowadays, most of the things are done through online. Most of the people do online shopping through any applications or websites rather than traditional shopping (Miguel et al., 2020).

Scammers Gautam Barad (19) of Amreli and Nilesh Babaria (21) of Jhunagadh have been arrested and booked under the provisions of the Information Technology (Amendment) Act and Indian Penal Code sections 120b (criminal conspiracy) and 419 (cheating by impersonation). These two suspects were reportedly nabbed by the Ahmedabad Cybercrime squad for stealing customers' login ids and banking information and then making purchases totaling many lakhs of rupees.

The accused had made many orders from well-known internet retailers. Meanwhile police had done inquiry and recovered more than 80 things including mobile phones, computer accessories, clothing, household items, designer shoes, digital gadgets and utensils amounting Rs.12 lakhs.

Cyber Crime Cell Deputy Commissioner Amit Vasava claims that two scammers downloaded a "coping file" from a telegram app using a hacking tool. Using a proxy IP address and remote desktop access tool, they used the consumers' login information to access the shopping websites and prepay for their purchases. The accused then attempted to alter the customer's order by altering their contact information, including their cell phone number and shipping address.

Case 3: Drugs Trafficking Through Dark Web and Sentenced to Over 11 Years of Prison

David Lee White, aged 52, of Chandler, Arizona, was sentenced on February 17,2022(Thursday) by U.S. District Judge Troy L. Nunley to 11 years and three months in prison (Stubbs-Richardson et al., 2018).

U.S. Attorney Phillip A. Talbert claims that White and two other defendants from Chandler, Arizona, Jason Keith Arnold and Alicia McCoy, respectively ran the vendor accounts SICKNESSVERSION and 23MIGHTYMOUSE on the Dream dark web marketplace. Through the dark web, these vendor accounts sold drugs like heroin, cocaine, methamphetamine, and marijuana that are illegal on a federal level.

The law enforcement agents starting working covertly purchased heroin and other controlled narcotics from these vendor accounts on various days in 2018, and had them sent to an address in the Eastern District of California. Agents immediately started executed a search order at a motel room in which White and McCoy were residing and seized at least 315 grams of heroin, 45 grams of cocaine, 593 grams of methamphetamine, and 30 grams of marijuana. The agents also found a Kel-tec 9 mm handgun loaded and unregistered in the motel room's nightstand.

After this incident, Arnold pleaded guilty on March 5, 2020, and also set to be sentenced on March 31, 2022. While charges are still pending against McCoy, and a status conference is planned for April 4, 2022. In this case, the product of an investigation by the Northern California Illicit Digital Economy (NCIDE) Task Force, which includes agents from Homeland Security Investigations, the Federal Bureau of Investigation, the United States Postal Inspection Service, the United States Postal Service Office of Inspector General, the Internal Revenue Service - Criminal Investigation, and the Drug Enforcement Administration. The NCIDE Task Force is a federal task force focused on targeting all sorts of unlawful dark web and cryptocurrency activities in the Eastern District of California and beyond. Assistant United States Attorneys Paul Hemesath and Sam Stefanki prosecuted the case.

Case 4: Washington Woman of Cyberstalking for Harassing and Threatening Montana Man and Public Officials

Based on the evidence presented by United States Attorney Leif M. Johnson to the court. On February 16, 2022, a jury found a lady from Washington guilty of cyberstalking a guy she met online in Montana, as well as state authorities, by hacking the man's email and sending harassing and threatening communications. On February 14, 2022, the trial got underway. After a three-day trial, a Walla Walla, Washington, jury found 60-year-old Kathy Ann Hendrickson, as Kathy Thorberg, guilty of cyberstalking per the indictment. Hendrickson could spend up to five years in jail, pay a $250,000 fine, and be placed on three years of probation (Liu & Young, 2018; Papakyriakopoulos & Zuckerman, 2021).

On May 26, 2022, U.S. District Judge Donald W. Molloy scheduled a sentence hearing. Hendrickson was held until additional hearings; the court will determine any sentence after taking into account the U.S. Sentencing Guidelines and other statutory circumstances.

In court, U.S. Attorney Johnson presented evidence that supported his claim that "Hendrickson stole the victim's personal information, hacked into his email account, and then used her cell phone to threaten and harass not only the victim but also public officials." Those who use anonymity online to harass and stalk children will face severe consequences. I appreciate the work of FBI Special Agent Timothy J. Racicot and the Office of the United States Attorney, where he served as an assistant.

The government produced evidence in court records and at trial suggesting that Hendrickson met the victim, who is named only as "John Doe," on a dating website in 2015 or 2016. After exchanging messages online, John Doe invited Jane Doe to his Trout Creek home for multiple in-person meetings. John Doe and Kathy Ann Hendrickson broke up after a certain amount of time. After they broke up, he began receiving threatening texts, prank calls, and emails, and his money was occasionally taken. John Doe was sure Hendrickson was attempting to take his wallet. In May of 2018, Hendrickson began sending threatening emails to John Doe using an anonymous account, saying things like "I'm going to bring you down." Hendrickson also reported being threatened by John Doe to the Sanders County Sheriff's Office.

The government further claimed that Hendrickson hacked into John Doe's email account using the password she obtained, wrote insulting messages to John Doe's friends, and made threats on herself. Henderson reported receiving these threatening texts and sought the Sanders County Sheriff's Department to protect him from John Doe. In addition, Hendrickson sent threatening emails through John Doe's account to two Sanders County commissioners and a receptionist. The messages all dealt with the theme of "Death," and included statements like "Your (sic) going to die." And the bullying kept on going. In January of this year, the authorities claimed that Hendrickson had broken into John Doe's email account and threatened then-Governor Steve Bullock of Montana. This could be interpreted to say, "You are the worst Governor for our state. To everyone's benefit, please just leave. It would probably be for the best if you were fired for good. One of the several firearms that I possess would suffice. Keep your eyes and ears open.

After concluding that Hendrickson was the prime suspect, police searched her house and took her phone. Forensic analysis of the phone uncovered evidence linking Hendrickson's surveillance of John Doe, such as the presence of an app that masks the phone number when making calls. John Doe and his associates were the targets of numerous hoax phone calls from unfamiliar numbers throughout the entire year of 2018. Hendrickson was accused by authorities of using a burner phone to send threatening emails from a fake John Doe account. Within a few days of the search of her home and the seizure of her actual phone, Hendrickson used the burner phone to telephone the U.S. Attorney's Office in Yakima, Washington, and the FBI office in Richland, Washington, and ask when her phone may be restored.

PREVENTION STEPS FOR AVOID CYBERCRIME

Prevention From Being Hacked (or Cybercrime)

1. Always use strong password and pin and try to keep changing passwords frequently so that hackers can face difficulty to access the password or pin and stored data.
2. Never share your bank account details information to any strangers.
3. Always keep social media private and don't give bank account details, more personal details in the social media. Also don't give any information via http sites. Share less in online.
4. Secure your devices which you are using mostly or not like smartphones, tablet, computer, etc. and always do system update also make sure use strong password on the lock screen as well.
5. Always protect your computer with security software. Hackers may try to track about your information through websites, IP address etc.
6. Always lock or logout(logoff) your devices like computers, smartphones, tablet when it is not in use.
7. Because hackers can easily connect to public Wi-Fi and monitor your every move, they are a security risk that you should avoid at all costs.
8. Always enable two factor authentication(2FA).
9. It may take little or more time always try to recognize the signs fake or malware infected websites, recognize phishing emails and sms. If you suspect a phishing or fake email or message(sms) block the sender as soon as possible and try to verify email with domain's name.
10. Always be vigilant while downloading anything from internet.
11. Always do transaction in limit and reconcile transactions regularly.
12. Always check download and folder file in computer, phone (or tablet). If you come across anything out of the ordinary that you didn't download yourself, it's best to do some research online to figure out what it is before erasing it.

Suppose You Had Faced a Cybercrime and Your Money was Stolen by Hacker (or we can Say Fraudster) From Your Bank Account What to Do?

As we know that there are some cybercrime cases in which money duped from internet (can be termed as cybercrime) are rarely solved by themselves, through police (Elezaj et al., 2019). If action is taken within 24 hours of the fraudulent transaction, there is a good likelihood that the money will be recovered. There are many steps in which it can recover:

1. Firstly, contact to the bank and block the account (credit card/debit card).
2. Then you should change your username, password, and answers to any security questions.
3. The next step is to make sure the hacker hasn't altered any of your contact information.
4. Please notify the appropriate bank personnel about the fraud.
5. Report a fraud case to local police station.
6. Hackers and cybercriminals can quickly begin making online purchases and money transfers after gaining access to a user's login name, password, and PIN. Before the victim cancels the card or changes their bank account information, they try to withdraw as much money as possible. Keep a close eye on the specifics of any suspicious transactions, and remember that the bank only gives you the date and time of each transaction.
7. The bank will readily block the transaction and refund the money, so no fraudster will want to do business with a stolen card or hacked bank account. This will guarantee the return of the funds to your account. But this can only happen if the fraudster hasn't already completed the transaction by delivering the items or processing the payment in full.
8. If you have given out your bank account information via email or text message, you should check to see if the goods have been delivered or if the transaction has been completed. It typically takes longer than a day.
9. Not every hacking victim will have the same success getting in touch with the perpetrator, but it never hurts to try.

It can be difficult to recover funds after they have been transferred to an individual through an online payment system. However, if the cybercriminal makes a purchase of goods or services online, the chances of the transaction being blocked and the money being recovered are much higher.

CONCLUSION

This study examines criminal activity and the associated feelings of insecurity to shed light on certain behavioural patterns relevant to quantitative social science. This research was conducted by contrasting real-world occurrences and reactions with those depicted on social media. Official data and victimization surveys provide an accurate representation of the crime situation in a country or city; the latter also provides quantitative measurements of the fear of crime.

REFERENCES

Elezaj, O., Yayilgan, S. Y., Kalemi, E., Wendelberg, L., Abomhara, M., & Ahmed, J. (2019, December). Towards designing a knowledge graph-based framework for investigating and preventing crime on online social networks. In *International Conference on e-Democracy* (pp. 181-195). Springer.

Hipp, J. R., Bates, C., Lichman, M., & Smyth, P. (2019). Using social media to measure temporal ambient population: Does it help explain local crime rates? *Justice Quarterly*, *36*(4), 718–748. doi:10.1080/07418825.2018.1445276

Kilgo, D. K., Harlow, S., García-Perdomo, V., & Salaverría, R. (2018). A new sensation? An international exploration of sensationalism and social media recommendations in online news publications. *Journalism*, *19*(11), 1497–1516. doi:10.1177/1464884916683549

Liu, S., & Young, S. D. (2018). A survey of social media data analysis for physical activity surveillance. *Journal of Forensic and Legal Medicine*, *57*, 33–36. doi:10.1016/j.jflm.2016.10.019 PMID:29801949

Malleson, N., & Andresen, M. A. (2015). The impact of using social media data in crime rate calculations: Shifting hot spots and changing spatial patterns. *Cartography and Geographic Information Science*, *42*(2), 112–121. doi:10.1080/15230406.2014.905756

Miguel, C. S., Morales, K., & Ynalvez, M. A. (2020). Online Victimization, Social Media Utilization, and Cyber Crime Prevention Measures. *Asia-Pacific Social Science Review*, *20*(4).

Milivojevic, S., & McGovern, A. (2014). The death of Jill Meagher: Crime and punishment on social media. *International Journal for Crime, Justice and Social Democracy*, *3*(3), 22-39.

Müller, K., & Schwarz, C. (2021). Fanning the flames of hate: Social media and hate crime. *Journal of the European Economic Association*, *19*(4), 2131–2167. doi:10.1093/jeea/jvaa045

Papakyriakopoulos, O., & Zuckerman, E. (2021, May). The Media During the Rise of Trump: Identity Politics, Immigration, "Mexican" Demonization and Hate-Crime. In *Proceedings of the International AAAI Conference on Web and Social Media* (Vol. 15, pp. 467-478). 10.1609/icwsm.v15i1.18076

Prieto Curiel, R., Cresci, S., Muntean, C. I., & Bishop, S. R. (2020). Crime and its fear in social media. *Palgrave Communications*, *6*(1), 1–12. doi:10.1057/s41599-020-0430-7

Ristea, A., Al Boni, M., Resch, B., Gerber, M. S., & Leitner, M. (2020). Spatial crime distribution and prediction for sporting events using social media. *International Journal of Geographical Information Science*, *34*(9), 1708–1739. doi:10.1080/13658816.2020.1719495 PMID:32939153

Stubbs-Richardson, M., Rader, N. E., & Cosby, A. G. (2018). Tweeting rape culture: Examining portrayals of victim blaming in discussions of sexual assault cases on Twitter. *Feminism & Psychology*, *28*(1), 90–108. doi:10.1177/0959353517715874

Walsh, J. P., & O'Connor, C. (2019). Social media and policing: A review of recent research. *Sociology Compass*, *13*(1), e12648. doi:10.1111/soc4.12648

Williams, M. L., Burnap, P., Javed, A., Liu, H., & Ozalp, S. (2020). Hate in the machine: Anti-Black and anti-Muslim social media posts as predictors of offline racially and religiously aggravated crime. *British Journal of Criminology*, *60*(1), 93–117. doi:10.1093/bjc/azz064

Yang, D., Heaney, T., Tonon, A., Wang, L., & Cudré-Mauroux, P. (2018). CrimeTelescope: Crime hotspot prediction based on urban and social media data fusion. *World Wide Web (Bussum)*, *21*(5), 1323–1347. doi:10.1007/s11280-017-0515-4

Chapter 10
Surveying the Role of Bamboo in Tourism Industry:
A Conceptual Analysis for Mitigating Climate Change Issues

Smitha Amarail Mohanan
https://orcid.org/0000-0002-9597-0311
Christ University, India

Suja John
Christ University, India

ABSTRACT

There can be no health or sustainable development without mental health underscores the inseparable link between the SDGs and mental health. Consequently, achieving the SDGs is a pathway to achieving mental health. Tourism stands as a significant contributor to positively impacting the mental health and well-being of communities. Bamboo, known as 'Green Gold' due to its carbon sequestration capabilities and other eco-friendly traits, aids in reducing carbon emissions and achieving the SDGs. The researcher has formulated a model aimed at achieving the SDGs while addressing climate change concerns and promoting well-being.

INTRODUCTION

Nature exposure significantly enhances mental well-being for numerous individuals. Forest therapy has gained considerable recognition as an innovative approach for stress relief and promoting health. The three-day session in a bamboo forest

DOI: 10.4018/979-8-3693-2177-5.ch010

notably enhanced the physiological well-being of university students (Zeng et al., 2020). Although there is a widespread need for mental health support, the existing responses are often insufficient and inadequate on a global scale (WHO, 2022). In one research study, 238 outdoor nature and adventure tourists from diverse age and demographic groups were interviewed at subtropical national parks in Queensland, Australia, to discern whether they visited parks due to their existing happiness or if their park visits contributed to their happiness. A substantial majority, amounting to 82%, indicated that they became happier because of their park visits. Among those surveyed, 87.5% mentioned experiencing short-term emotional benefits, 60% reported medium-term stress relief, and 20% noted long-term shifts in their worldview (Buckley,2020). It consequently opens a substantial potential market for nature-based commercial tourism, particularly in the role of offering nature-based therapies, notably within national parks (Buckley,2020).

Engaging in leisure tourism is a voluntary pursuit aimed at enhancing personal well-being. According to Buckley (2020), commercial outdoor tourism ventures can notably enhance the well-being of women and families. Therefore, there is a correlation between mental well-being and tourism. Concerns about climate change affect both the well-being of individuals and the places they travel to. Becken (2013), in his review, asserts that the convergence of tourism and climate change has emerged as an independent and specialized field of study.

As tourism serves as a primary means for relaxation and fostering a healthy mental state, the surge in tourist numbers visiting certain destinations has resulted in the degradation of these places (Butler, 2006; Daly et al., 2020). The phenomenon of over-tourism significantly contributes to the problem of climate change. Climatic change and pollution are listed among the top 10 global crises, creating an alarming situation that needs to be resolved at the earliest. The widespread contribution of greenhouse gas emissions to climate change has prompted extensive research efforts to address this crisis (Turner, 2018). Whether assessed objectively or subjectively, pollution can markedly discourage prospective international tourists. (Xu & Reed, 2018) Climate preservation, environmental conservation, and natural resources would pave the way for an economic, environmental, and socially sustainable field of tourism (Lu et al., 2021). Research on climate and tourism has significantly grown over the last few decades (Pang et al., 2012). The contribution of the global carbon footprint caused by tourism has increased from 3.9 to 4.5 Gt CO2e (Lenzen (2018), contributing to 8% of the global greenhouse gas emissions. Research conducted in 58 major cities in China identified that atmospheric pollution could adversely affect the tourism economy of a city (Zhang et al., 2020). Marine wastes caused by tourism adversely affect the aquatic fauna and flora, especially waste, with low-value, nonrecyclable, non-biodegradable plastics, such as grocery bags, ropes, and forms, including marine debris. Chen (2019), in his research, shows how Xiamen

Bay, a tourist city, is affected by the marine plastic waste, which adversely affects the aquatic life of the area as well as the surrounding community. Souvenirs have an unavoidable relationship with tourism (Gordon, 1986). They are also a factor that leads to waste generation.

Cruise tourism also generates massive amounts of waste; for instance, one person on a cruise produces about 206 to 305 kgs of waste a day, which is more than the average production on land (Deepika Manoharan, 2021). Cruise wastes are often dumped into the ocean due to the absence of waste disposal procedures (Deepika Manoharan, 2021). Seasonal haze disaster is another contributor to pollution that the tourism industry faces. Followed by ethnographic research on urban practitioners, social media reports and commentaries state that geopolitical discussions of transboundary haze production are shaped by tourists and the tourism industry (Mostafanezhad & Evrard, 2018).

A lot of research has been conducted on the benefits and significance of bamboo. Here, the researcher tries to combine the significant findings on bamboo from previous literature and puts them under one umbrella. After that, the researcher explores the possibilities of incorporating bamboo into the tourism industry to reduce climate change issues. By involving community participation through bamboo and getting them involved in the tourism industry, the researcher tries to identify the possibilities of attaining Sustainable Development Goals (SDGs) through this paper. Here, the researcher tries to incorporate the significant features of bamboo, connect them with the tourism industry, and explore the possibilities to attain SDGs.Climate change represents a worldwide challenge that has spurred extensive research across various academic disciplines. However, the tourism industry's role in this issue has received comparatively less attention (Buckley, 2011). The global crises of climatic change and pollution rank among the top 10, necessitating urgent resolution. Greenhouse gas emissions significantly contribute to climate change, driving extensive ongoing research to address this crisis (Turner, 2018). Forging a path towards an economically, environmentally, and socially sustainable tourism sector involves prioritizing climate preservation, environmental conservation, and prudent use of natural resources (Lu et al., 2021). Over recent decades, research at the intersection of climate and tourism has seen substantial growth (Pang et al., 2012). The tourism industry's contribution to the global carbon footprint has risen from 3.9 to 4.5 Gt CO2e (Lenzen, 2018), constituting 8% of global greenhouse gas emissions. Research conducted in 58 major Chinese cities has highlighted the adverse impact of atmospheric pollution on a city's tourism economy (Zhang et al., 2020). Tourism-generated marine waste negatively affects aquatic ecosystems, particularly nonrecyclable, non-biodegradable plastics like grocery bags, ropes, and forms, collectively categorized as marine debris (Chen, 2019). Chen's (2019) study delves into the detrimental effects of marine plastic waste on aquatic life and local communities in Xiamen Bay, a tourist city. Cruise tourism, in particular, generates substantial waste, with an individual on a cruise producing 206 to 305 kg of waste per day, surpassing average land-based waste production (Deepika Manoharan, 2021). Due to inadequate waste disposal procedures, cruise waste often finds its way into the ocean (Deepika Manoharan, 2021). Another contributor to pollution faced by the tourism industry is seasonal haze disasters. Research indicates that discussions regarding transboundary haze production are influenced by tourists and the tourism industry, as evidenced by ethnographic research on urban practitioners, social media reports, and commentaries (Mostafanezhad & Evrard, 2018).

Extensive research has explored the benefits and significance of bamboo. This paper consolidates significant findings from previous literature on bamboo and examines the potential integration of bamboo to mitigate climate change. The researcher explores community engagement through bamboo, aiming to involve local communities in the tourism industry and identify opportunities to achieve Sustainable Development Goals (SDGs). Thus, community participation leads to benefits for the mental health of individuals. By aligning the salient features of bamboo with the tourism sector, this research endeavors to uncover pathways towards attaining the SDGs. "There can be no health or sustainable development without mental health" by Goodwin and Zaman (2023) states that the SDGs and mental health are mutually connected. Hence, the attainment of SDGs leads to the attainment of mental health.Goodwin and Zaman (2023) say, "The presence of mental health is imperative for both health and sustainable development," highlighting the interconnectedness between mental health and the SDGs. Therefore, achieving the SDGs leads to achieving mental health.

PROBLEM STATEMENT

Numerous factors contribute to climate change problems, and this discussion focuses on the tourism sector as one of the key contributors. The tourism industry significantly contributes to environmental pollution, particularly playing a noteworthy role in climate change (Zhang et al., 2020). Multiple studies have identified air transportation, accommodation, and other segments of the tourism economy as significant sources of greenhouse gas (GHG) emissions (Lenzen et al., 2018; Scott et al., 2010; Haites, 2011). During the peak seasons at Santa Marta and the beaches along the Colombian Caribbean, there is an observed increase in pollution levels compared to the off-peak tourist season. Tourism plays a significant role in contributing to litter pollution, with plastics making up a substantial portion, ranging from 30% to 77% of the larger debris, while microplastics are found at levels ranging from 1 to 355 items per square meter. Consequently, it is imperative that we take immediate action to safeguard our environment (Garcés-Ordóñez et al., 2020).

Scholars addressing concerns related to over-tourism emphasize the essential need for regulatory measures and government guidance. (Koens et al., 2018). Researchers are actively seeking ways to mitigate GHG emissions. They propose diverse biotic and abiotic carbon sequestration methods to decrease atmospheric carbon dioxide (Lal, 2007). This study aims to explore the potential of 'Bamboo - The Green Gold' in climate resilience, promoting sustainable, environmentally friendly tourism and developing a healthy community.

METHODOLOGY

The method used for this conceptual paper is based on literature reviews from peer-reviewed journal articles based on the inclusion criteria. A set of keywords were selected and cross-referenced with papers from the tourism industry. The articles were extracted from Scopus-indexed journals using the keywords: benefits of bamboo, sustainable development, environmental impact, community development, and mental well-being.

INTERLINKING TOURISM, WELL-BEING, AND CLIMATE CHANGE

Fostering partnerships among government agencies, environmental organizations, local communities, and tourist stakeholders is essential to the successful implementation of Nature-based Solutions (NbS) and Ecosystem-based Adaptation (EbA). The integration of NbS and EbA into tourism and well-being strategies is crucial for enhancing climate change resilience. This all-encompassing strategy recognizes the complex relationship that exists between environmental health and human well-being, highlighting the necessity of protecting both concurrently. As a result, this integration strengthens efforts to combat climate change, improves tourist well-being, and increases the resilience of local communities (Loehr et al., 2022).

NbS and EbA strive to tackle societal issues by harnessing the advantages provided by ecosystem services for the benefit of the human population and its well-being. In their research, Loehr et al. (2022) utilize a figure adapted from the IUCN NbS framework (Cohen-Shacham et al., 2016) to illustrate how EbA can be employed as NbS for mitigating climate change and promoting well-being.

From the Figure 1, we can understand that EbA ensures the availability of essential ecosystem services, including freshwater quality, food, soil fertility, shelter, flood risk reduction, and the provision of other crucial natural resources necessary for sustaining life, particularly among impoverished individuals residing in rural regions

Figure 1. Ecosystem-based adaptation Nbs to address climate change and create well-being
Note: IUCN = International Union for Conservation of Nature; NbS = Nature-based Solutions.

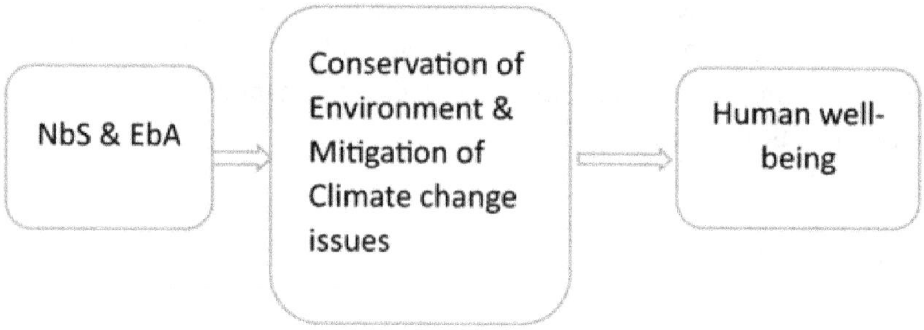

(Munang et al., 2011). Consequently, EbA has the potential not only to mitigate climate-related risks but also to enhance overall human well-being.

In this context, we explore the potential of bamboo, a renewable and biodegradable resource, as an Ecosystem-based Adaptation (EbA) strategy. We begin by examining the advantages of bamboo and its integration into the tourism industry to address climate change challenges and promote the well-being of individuals.

DISCUSSIONS

Benefits of Bamboo

The world is grappling with several global crises, including high population growth, diverse climate patterns, and water scarcity (Duda & El-Ashry, 2000; Vorosmarty et al., 2000). Many countries have initiated the development of sustainable green technologies to address these challenges. It is crucial to underscore the importance of implementing effective planning strategies to reduce adverse effects on the physical environment. Additionally, it is essential to recognize the substantial disparities in impact between historic districts and emerging developments (Vehbi & Doratli, 2010). Bamboo, renowned as the best carbon absorber to mitigate anthropogenic carbon emissions, plays a significant role in global climate change mitigation. It functions as an atmospheric air purifier, with a mature bamboo tree capable of producing up to 300 kilograms of oxygen annually, enough to sustain a person for a year (Sundararaju, 2021). Bamboo is also exceptionally fast-growing and possesses a lengthy lifespan (Sundararaju, 2021; Cho et al., 2011), making it a potent carbon dioxide sequestration tool, with one acre of bamboo cultivation capable of capturing

80 tons of CO2 annually (Sundararaju, 2021; Emamverdian et al., 2020). Bamboo's remarkable properties for carbon sequestration are globally relevant, except in Europe and Antarctica; bamboo grows on every continent (Lessard & Chouinard, 1980).

Additionally, bamboo is rich in proteins, vitamins, carbohydrates, dietary fibers, and antioxidants (Luo et al., 2017; Zheng et al., 2014). Its edible shoots are protein-rich and nutrient-dense, making bamboo shoots a highly sought-after commodity in the international market due to their increasing consumption and demand (Dwivedi et al., 2019). Bamboo cultivation can provide farmers with an additional income source, doubling their economic empowerment (Dwivedi et al., 2019). Bamboo exhibits a greater capacity for storing water (Dierick et al., 2010). Bamboo shoots also hold medicinal value, with properties like reducing hyperthyroidism and combating the proliferation of cancer cells (Bajaj et al., 2016; Wang et al., 2021). Furthermore, bamboo offers a wide array of health benefits, including anti-cancer, antioxidant, anti-aging, weight loss, probiotic, and cardioprotective properties (Behera & Balaji, 2021). In India, fermented bamboo shoots are referred to as "Green Gold" (Behera & Balaji, 2021). Bamboo juice extracted from cooked leaves possesses high medicinal value, with flavonoids from the leaves countering free radicals, lipids, and cholesterol, providing anti-aging, antifatigue, and skin moisturizing benefits (Wang et al., 2021). Fermented bamboo has gained worldwide popularity due to its antibacterial characteristics and medicinal properties, making it a highly sought-after product. However, industry stakeholders have yet to provide scientific evidence to substantiate these claims (Afrin et al., 2012).

The world is home to 1,250 bamboo species within 75 genera of woody bamboos (Benton, 2015). These species can be categorized into three groups based on their suitability for various uses such as manufacturing, farming, decoration, and environmental construction (Maoyi & Banik, 1996; (Wysocki et al., 2015) Bamboo species exhibit different growth habits, including clumping (sympodial bamboos), which spread slowly to form clusters, and species with longitudinal rhizomes (monopodial bamboos), which are versatile and widely cultivated globally. Some species combine characteristics of both, known as amphipodial bamboos, possessing a mix of cluster and longitudinal rhizomes (Ruiz-Pérez et al., 2001). The manufacturing industry finds various applications for bamboo fibers and craft items (Abdul Khalil et al., 2012; Wang et al., 2021), including tools for agriculture, fishing, handicrafts, musical instruments, furniture, and laminated flooring (Scurlock et al., 2000). Furniture and handicrafts are particularly sought-after commodities in the international market, and China boasts extensive bamboo forests covering over 300,000 hectares (Wang et al., 2021).

Bamboo also plays a pivotal role in biomass energy (Scurlock et al., 2000; Sharma et al., 2018). Advancements in bamboo's properties for fuel development could significantly reduce global pollution caused by the transportation sector. In India, the paper industry relies on two-thirds of its pulp production from bamboo (Dwivedi et al., 2019), reducing dependence on wood pulp and helping combat deforestation, thereby improving the climate.

Research indicates that bamboo-based buildings have a lower environmental impact compared to traditional structures. Bamboo fibers exhibit excellent mechanical properties suitable for construction, and bamboo in both dry and boiled states possesses properties similar to granite and concrete (Yap & Yahya, 2016). Embracing industrialized bamboo production for construction purposes would reduce the demand for wood production and mitigate deforestation. Initiatives in Africa are actively cultivating, managing, and utilizing Ethiopian bamboo species to combat deforestation in East Africa by promoting bamboo as a sustainable wood source and a basis for profitable businesses (Böck, 2014). Bamboo cultivation and processing hold immense potential for global climate change mitigation.

In Ethiopia, bamboo offers an ideal source of locally sustainable engineered materials for rapidly growing urban areas across Africa (Böck, 2014). Public-private partnerships have developed sustainable business models to promote bamboo as a construction raw material. However, bamboo-based buildings must be carefully managed at their end-of-life stages to efficiently store carbon dioxide (CO2) (Zea Escamilla et al.). Bamboo-based porous activated carbons exhibit promising textural features for CO2 capture using phosphoric acid (H3PO4) (Syafiqah Ismail et al., 2022).

With its strong compressive characteristics in both dry and boiled states, comparable to granite and concrete (Israr et al., 2016), bamboo has the potential to replace energy-intensive, unsustainable, and costly materials like steel and concrete, a prospect currently under research and development (Archila et al., 2018). Bamboo emerges as a primary ingredient for environmentally friendly constructions and interior designs, offering a renewable, biodegradable, and environmentally sustainable alternative to non-biodegradable products such as steel and concrete, thereby earning its moniker as “Green Gold Bamboo.”

Carbon Sequestration Capacity of Bamboo

Bamboo’s purported capability for carbon sequestration is one of the most important factors that increase the demand for this plant. Bamboos are effective carbon dioxide absorbers in both above-ground carbon (AGC) and below-ground carbon (BGC) in roots and rhizomes. Earlier research studies revealed that the bamboo can absorb as much as 12 tons of CO2 per hectare per year, giving it a promising and vital role in stabilizing the environment. A mature grove of bamboo generates 30% to 35% more oxygen than an equal area of forest trees. In order to produce it, bamboo absorbs the carbon dioxide from the atmosphere and stores much of it in its roots and biomass.

Carbon sequestration is probably its most significant contribution to the battle against global warming and is determined by the news columns produced annually. The development of bamboo cultivation and its management have social, ecological, and economic benefits and promote an effective solution for climate change mitigation strategy (Jyoti Nath & Das, 2012). The total carbon storage in the soil contributed 84.6% of the total (50.1 Mg ha21), followed by carbon in above-ground vegetation, which produced 15% (9 Mg ha21), and carbon in litter floor mass contributed 0.4% (0.2 Mg ha21) (Jyoti Nath & Das, 2012).

Benefits of Bamboo in Community Development and Well Being

In light of the increasing concern about the adverse impacts of mass tourism and the heightened enthusiasm among travelers for natural settings and authentic experiences, there has been a notable surge in the adoption of Community-Based Ecotourism (CBET) and Ecotourism, both of which emphasize sustainability. The ecotourism framework proposed by Ross and Wall (1999) asserts that successful ecotourism relies on a positive synergy among people, resources, and tourism, wherein each element contributes positively to the others (Ross & Wall, 1999). In African protected ecotourism areas, ecotourism initiatives safeguard both the natural environments and the adjacent rural communities. The monthly income generated from employment in ecotourism has been demonstrated to empower households, allowing them to invest in assets and education (Snyman, 2012). This has led to enhanced financial security and improved social welfare in isolated rural areas. In CBET, a key emphasis is placed on community involvement and well-being over other factors (Ait-Yahia Ghidouche & Ghidouche, 2019).

The development of the bamboo industry not only benefited the environment but also increased employment opportunities (Wang et al., 2021), creating good livelihoods and significantly reducing climate change issues (Jyoti Nath & Das, 2012). Bamboo's high elasticity feature is appropriate for building materials in earthquake-prone areas. Research shows that wood-frame construction could withstand the impact of even large earthquakes without much damage (Rainer & Karacabeyli 1999). Using bamboo for environment-friendly construction projects would help to reduce poverty in mountainous regions. Apart from the economic growth, developing the bamboo industry would help to protect the national culture and develop ecotourism and a community-based ecotourism landscape (Wang et al., 2021).

The exciting point about bamboo is that it grows in all continents except Europe and Antarctica (Chouinard, 1980). As bamboo is a renewable energy source, its speedy growth will help to use a mediating variable to attain SDGs and enhance the livelihoods of communities participating in its product activities.

Bamboo and Tourism Industry

Bamboo offers numerous benefits that can elevate opportunities within the tourism industry. Implementing these practices can foster community development, mental health, and overall well-being.

1. Bamboo farmers can form a cooperative society, which would help to sustain the market prices, avoid hostile competition and improve economic growth (Wang et al., 2021). In ecotourism, bamboo has a significant commercial value. The panoramic green visuals of its forests attract ecotourists. Ecotourism is a remedy to reduce the negative effects of mass tourism. Promoting the bamboo forests in ecotourism destinations will promote community-based ecotourism and lead to local development through the marketing of handicrafts. Promoting ecotourism would reduce poverty and help to uplift the community (Scheyvens, 1999).
2. The restaurants and huts using bamboos near historic destinations would balance the carrying capacity of tourists and would not affect the destination's life cycles. The construction could be encouraged with bamboo products to conserve the environment. Bamboo tree huts and cottages with five-star interior facilities would attract domestic and international tourists. Furniture and interior decorations made of bamboo would also help to reduce carbon emissions. Harmful paints can be avoided by using natural bamboo. That would be a unique experience for tourists. These accommodation facilities would help to give relief from monotony to the international tourist.
3. As tourists transition into active participants and co-creators of souvenirs, they open new possibilities for forming a deeper connection with the souvenir, as it becomes something they have actively contributed to crafting (Anastasiadou & Vettese, 2018). Tourism destinations can encourage souvenirs made with biodegradable bamboo frames that would not generate waste. Introducing background pictures of the destinations fixed on frames of bamboo would attract tourists, taking away the feel and essence of visited destinations.
4. Forks and spoons, unavoidable at restaurants, can be replaced with bamboo-forks, spoons, plates, and bowls. This would reduce the usage of plastics and steel. It would generate job opportunities for the local community. Files, pens, bags, and other stationery materials can replace bamboo pens, files, and bags.
5. Cosmetics made with bamboo, especially anti-aging creams, are in high demand internationally.
6. Tourist boats are the main threat to the aquatic fauna and flora. They could be replaced by bamboo coracles, boats, and bridges across small streams. They would reduce air and water pollution and conserve aquatic plants and animals.

7. Bamboo floorings must be encouraged in tourist hotspots, providing wooden finishings. The homestays and cottages could be encouraged to make bamboo constructions. Its antifungal ability would help to fight harmful insects.
8. Bamboo food items are delicious and healthy. The bamboo shoots and other products can be incorporated into restaurants as an important dish. Bamboo juice and beer, too, are healthy and tasty.
9. The bamboo pulp can be used in the paper industry, reducing deforestation and mitigating climate change issues.
10. Bamboo gardens, apart from benefiting rural life, can also reduce climate issues. Bamboo gardens can also be used at tourist destinations.

BAMBOO: A BRIDGE TO UN-SUSTAINABLE DEVELOPMENT GOALS AND MENTAL HEALTH

The significance of mental health within SDG 3, centered on promoting good health and well-being, is intrinsically linked to various other SDGs. This is due to the correlation between mental health and poverty (SDG 1), as well as the endeavor to diminish inequalities (SDG 10), among other SDGs. Consequently, adequately addressing the UN Sustainable Development Goals is notably challenging without prioritizing mental health (Goodwin and Zaman, 2023). When compared to physical health, the international attention given to mental health is significantly lacking in terms of funding, education, and implementation (World Health Organization, 2020). This gap can be bridged by attainment of SDG's trough incorporating bamboo products in tourism industry.

A targeted initiative aimed at nurturing and overseeing bamboo growth in ecotourism hotspots could significantly contribute to the realization of the United Nations' Sustainable Development Goals (SDGs). An alternative and sustainable income source for rural communities could be established by commercializing bamboo products in tourist destinations. These endeavors would actively promote the goals of eradicating hunger, improving health, and fostering prosperity.

Bamboo possesses a multitude of applications that seamlessly align with the 17 SDGs. The initial objective, "No Poverty," could be effectively addressed by creating employment opportunities and implementing bamboo-related constructions, products, and tourism initiatives (Phi et al., 2018; Zapata et al., 2011). The subsequent goal, "Zero Hunger," could be resolved by promoting job opportunities and implementing alternative tourism approaches like ecotourism, community-based ecotourism, responsible, and green tourism. This approach would enable communities to engage in tourism activities and alleviate poverty. Furthermore, bamboo cultivation contributes to air purification and creates a healthier environment, aligning with the third SDG, "Good Health and Well-being."

Enhancing the quality of tourism education is vital in meeting industry needs, identifying gaps, and harmonizing education efforts (Seraphin et al., 2020). Self-help groups and bamboo corporations can collaborate to enhance education quality for rural vendors, making them self-sufficient in tourism activities without intermediaries. Gender equality, the fifth SDG goal, emphasizes equal job opportunities and women's empowerment, training, participation, and employment in tourism (Rasoolimanesh et al., 2020). Manufacturing bamboo products can empower women economically and reduce gender inequality.

The sixth SDG goal, "Clean Water and Sanitation," benefits from bamboo cultivation by preserving soil moisture and ensuring access to pure drinking water. Bamboo also reduces air pollution, releasing excess oxygen and contributing to sustainable renewable energy. Research into bamboo-based biodiesel production is ongoing. Goal 8, "Decent Work and Economic Growth," highlights the importance of collaborative economic efforts (Gössling & Hall, 2019). Bamboo-based job opportunities would benefit the entire community. Goal 9, "Industry, Innovation, and Infrastructure," can be advanced through tourism activities (Gössling & Hall, 2019) and bamboo projects.

Reducing inequality, the tenth SDG, is closely linked to ensuring equal participation of residents and non-residents in tourism. Sustainable job creation would decrease social inequalities, ensuring that everyone can earn and equitably share the benefits derived from bamboo and the tourism industry (Rasoolimanesh et al., 2020).

The eleventh SDG emphasizes "Sustainable Cities and Communities." Ecotourism in bamboo-rich regions can enhance urban and community sustainability. Promoting bamboo cultivation and production through green technology in green tourism reduces reliance on natural resources and supports conservation, thereby achieving the twelfth SDG goal of "Responsible Consumption and Production." Goal 13 urges action on climate change, which bamboo cultivation can enable by mitigating global warming and air pollution. Bamboo products, being biodegradable, reduce plastic waste in water bodies, contributing to the fourteenth SDG, "Life Below Water." Cultivating bamboo along riverbanks can reduce flood risks. "Life on Land" is threatened by factors like deforestation and land degradation, which bamboo products and cultivation can mitigate by reducing dependence on other trees and improving soil quality, addressing the fifteenth SDG goal.

Bamboo businesses bolster communities and reduce crime and violence, creating a peaceful environment in alignment with the sixteenth SDG, "Peace and Justice through Strong Institutions." Trade and export businesses centered around bamboo can facilitate global partnerships and contribute to achieving the seventeenth SDG goal of "Partnerships for the Goals."

Figure 2 shows the relationship between SDGs and their achievements through bamboo products. (Source -Author's representation)

Figure 2.

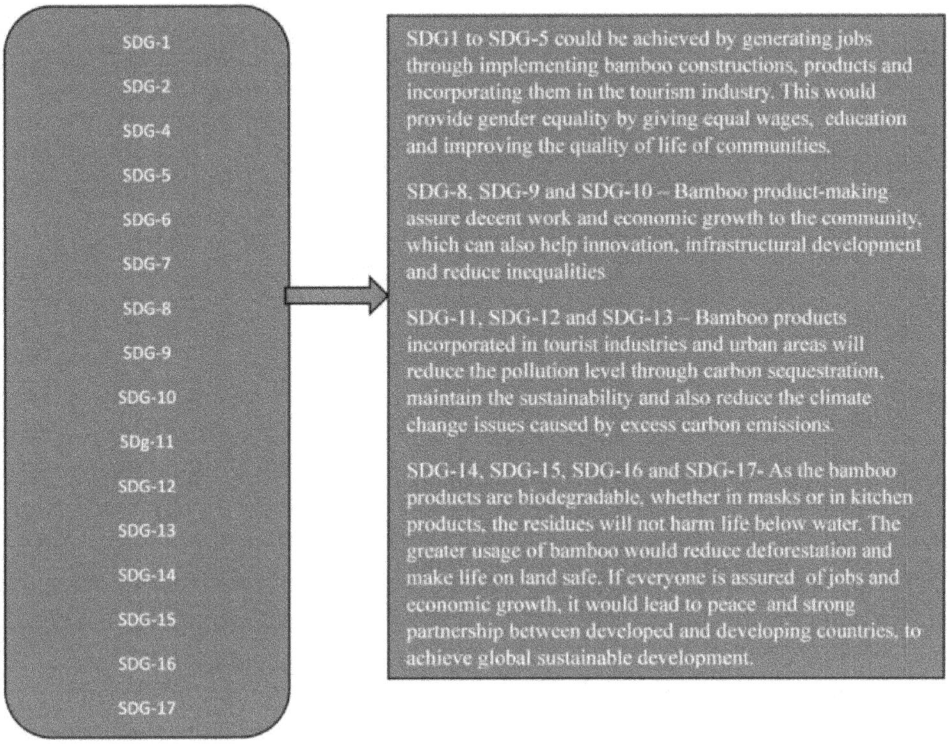

The Figure 3 shows that the researcher tries to incorporate the above possibilities of the tourism industry to incorporate bamboo, the Green Gold, which leads to attaining SDGs and mental health. Here, both the cultivation and products of Bamboo and SDGs stand as mediating variables.

The incorporation of bamboo products will increase the possibility of income and infrastructure also the well-being and mental heal of the individual globally. The integration of bamboo products will enhance income opportunities, infrastructure development, as well as the overall well-being and mental health of individuals worldwide. Reducing harmful pollutants and energy consumption has the potential to mitigate climate change and pollution, thereby preserving the environment. Bamboo products can serve as alternatives to hazardous substances in this effort.

Figure 3.

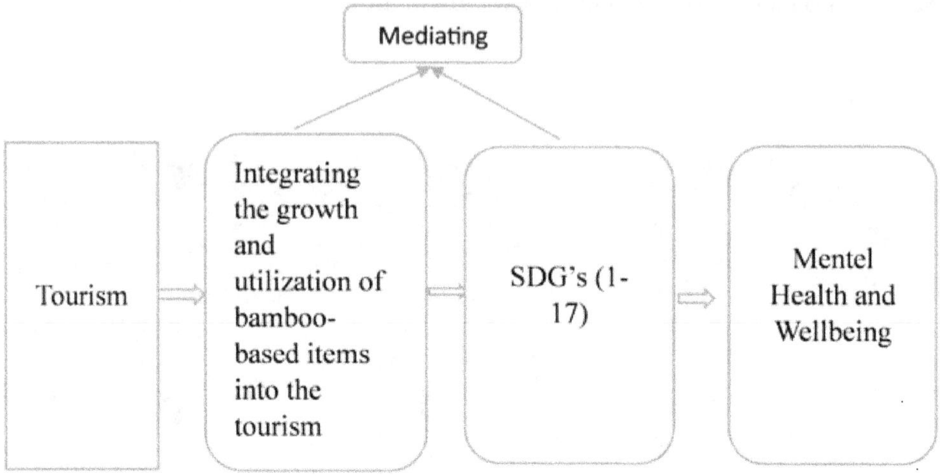

CONCLUSION

The researcher proposes the potential integration of innovative strategies to achieve the 17 Sustainable Development Goals (SDGs). As illustrated in the model, incorporating renewable resources such as bamboo into the tourism sector can create employment opportunities for the local populace, while simultaneously preserving and balancing the environment, thus mitigating climate change concerns. The integration of bamboo initiatives and materials within the tourism industry is poised to facilitate the expansion of the green economy. Recent research underscores that natural surroundings can enhance both mental well-being and physical health.

Future researchers have the opportunity to comprehensively explore the array of bamboo products and their integration into the tourism sector, examining their viability and practicality. While the model demonstrates the potential for the tourism industry, bolstered by bamboo integration, to encompass the entire spectrum of SDGs, it is essential to empirically test this practicability. Additionally, emphasis should be placed on research focusing on how bamboo products can be effectively utilized to benefit communities.

Ecotourism and community-based tourism present viable approaches for achieving the UN SDGs and uplifting communities. Notably, bamboo products are biodegradable; bamboo waste naturally decomposes in the soil, and bamboo charcoal can serve as a fertilizer. This study aims to establish a strong connection between the Sustainable Development Goals and Mentel health in connection with tourism industry. Countless innovative ideas and opportunities for fostering a sustainable economy remain to be explored comprehensively.

LIMITATIONS OF THE STUDY

The researcher exclusively relied on literature reviews, with a particular emphasis on highlighting the advantages of bamboo. However, it's important to note that the researcher's proposed model has not undergone any practical testing. To bridge the gap between theory and real-world application, empirical research is essential. This involves gathering empirical data to validate the action-oriented projects proposed. Furthermore, the model itself has not been put to the test. Researchers have the opportunity to delve deeper into the potential of utilizing bamboo, often referred to as "Green Gold," within the tourism industry as a means of achieving Sustainable Development Goals (SDGs). This can be accomplished by linking bamboo products to tourism initiatives aimed at attaining SDGs, a strategy that can be tailored to countries worldwide.

The model's applicability holds significant promise, considering that bamboo is naturally abundant on all continents except Europe and Antarctica. It's worth noting that bamboo has even been introduced in Europe (as documented by Benton in 2015). This study is expected to be a valuable resource for researchers focusing on SDGs within the context of the tourism industry.

REFERENCES

Abdul Khalil, H. P. S., Bhat, I. U. H., Jawaid, M., Zaidon, A., Hermawan, D., & Hadi, Y. S. (2012). Bamboo fibre reinforced biocomposites: A Review. *Materials &amp. Materials & Design*, *42*, 353–368. doi:10.1016/j.matdes.2012.06.015

Afrin, T., Tsuzuki, T., Kanwar, R. K., & Wang, X. (2012). The origin of the antibacterial property of bamboo. *Journal of the Textile Institute*, *103*(8), 844–849. doi:10.1080/00405000.2011.614742

Ait-Yahia Ghidouche, K., & Ghidouche, F. (2019). Community-based ecotourism for preventing overtourism and tourism phobia: Algerian associations' viewpoints. *Worldwide Hospitality and Tourism Themes*, *11*(5), 516–531. doi:10.1108/WHATT-06-2019-0035

Anastasiadou, C., & Vettese, S. (2018). Digital revolution or plastic gimmick? Authenticity in 3D souvenirs. In Tourism Social Science Series (Vol. 24). doi:10.1108/S1571-504320180000024011

Archila, H., Kaminski, S., Trujillo, D., Zea Escamilla, E., & Harries, K. A. (2018). Bamboo reinforced concrete: A critical review. *Materials and Structures*, *51*(4), 102. Advance online publication. doi:10.1617/s11527-018-1228-6

Becken, S. (2013). A review of tourism and climate change as an evolving knowledge domain. *Tourism Management Perspectives*, *6*, 53–62. doi:10.1016/j.tmp.2012.11.006

Behera, P., & Balaji, S. (2021, June 1). Health benefits of fermented bamboo shoots: The twenty-first century green gold of Northeast India. *Applied Biochemistry and Biotechnology*, *193*(6), 1800–1812. doi:10.1007/s12010-021-03506-y PMID:33496924

Benton, A. (2015). *Priority species of bamboo*. doi:10.1007/978-3-319-14133-6_2

Böck, F. (2014). Green gold of Africa–Can growing native Bamboo in Ethiopia become a commercially viable business? *Forestry Chronicle*, *90*(5), 628–635. doi:10.5558/tfc2014-127

Buckley, R. (2011). Tourism and environment. *Annual Review of Environment and Resources*, *36*(1), 397–416. doi:10.1146/annurev-environ-041210-132637

Buckley, R. (2020). Nature tourism and mental health: Parks, happiness, and causation. *Journal of Sustainable Tourism*, *28*(9), 1409–1424. doi:10.1080/09669582.2020.1742725

Buckley, R., & Westaway, D. (2020). Mental health rescue effects of women's outdoor tourism: A role in COVID-19 recovery. *Annals of Tourism Research*, *85*, 103041. doi:10.1016/j.annals.2020.103041 PMID:33100433

Butler, R. W. (2006). *The Tourism Area Life Cycle*. Academic Press.

Chen, C., Chitose, A., Kusadokoro, M., Nie, H., Xu, W., Yang, F., & Yang, S. (2021). Sustainability and challenges in biodiesel production from waste cooking oil: An advanced bibliometric analysis. *Energy Reports*, *7*, 4022–4034. doi:10.1016/j.egyr.2021.06.084

Chen, H., Wang, S., Guo, H., Lin, H., Zhang, Y., Long, Z., & Huang, H. (2019). Study of marine debris around a tourist city in East China: Implication for waste management. *The Science of the Total Environment*, *676*, 278–289. doi:10.1016/j.scitotenv.2019.04.335 PMID:31048159

Cho, E., Um, Y., Yoo, S. K., Lee, H., Kim, H. B., Koh, S., Shin, H. C., & Lee, Y. (2011). An expressed sequence tag analysis for the fast-growing shoots of bambusa edulis murno. *Journal of Plant Biology*, *54*(6), 402–408. doi:10.1007/s12374-011-9179-2

Daly, P., Dias, A. L., & Patuleia, M. (2020). The impacts of tourism on Cultural Identity on lisbon historic neighbourhoods. *Journal of Ethnic and Cultural Studies*, *8*(1), 1–25. doi:10.29333/ejecs/516

Deepika Manoharan, S. J. (2021, March 17). *Cruise tourism in India: Sailing into troubled waters*. Retrieved from Down To Earth. https://www.downtoearth.org.in/blog/climate-change/cruise-tourism-in-india-sailing-into-troubled-waters-75995

Dierick, D., Hölscher, D., & Schwendenmann, L. (2010). Water use characteristics of a bamboo species (Bambusa Blumeana) in the Philippines. *Agricultural and Forest Meteorology*, *150*(12), 1568–1578. doi:10.1016/j.agrformet.2010.08.006

Duda, A. M., & El-Ashry, M. T. (2000). Addressing the global water and environment crises through integrated approaches to the management of land, water and ecological resources. *Water International*, *25*(1), 115–126. doi:10.1080/02508060008686803

Dwivedi, A. K., Kumar, A., Baredar, P., & Prakash, O. (2019). Bamboo as a complementary crop to address climate change and livelihoods – Insights from India. *Forest Policy and Economics*, *102*(February), 66–74. doi:10.1016/j.forpol.2019.02.007

Emamverdian, A., Ding, Y., Ranaei, F., & Ahmad, Z. (2020). Application of bamboo plants in nine aspects. *TheScientificWorldJournal*, *2020*, 1–9. doi:10.1155/2020/7284203 PMID:33061861

Garcés-Ordóñez, O., Espinosa Díaz, L. F., Pereira Cardoso, R., & Costa Muniz, M. (2020). The impact of tourism on marine litter pollution on Santa Marta beaches, Colombian Caribbean. *Marine Pollution Bulletin*, *160*, 111558. Advance online publication. doi:10.1016/j.marpolbul.2020.111558 PMID:32853837

Goodwin, H. (2017). *The challenge of overtourism*. Responsible Tourism Partnership Working Paper 4. October.

Goodwin, J., & Zaman, U. (2023). Mental health stigma and UN Sustainable Development Goals. *Frontiers in Psychiatry*, *14*, 1190406. doi:10.3389/fpsyt.2023.1190406 PMID:37091707

Gordon, B. (1986). The souvenir: Messenger of the extraordinary. *Journal of Popular Culture*, *20*(3), 135–146. doi:10.1111/j.0022-3840.1986.2003_135.x

Gössling, S., & Michael Hall, C. (2019). Sharing versus collaborative economy: How to align ICT developments and the SDGs in tourism? *Journal of Sustainable Tourism, 27*(1), 74–96. Advance online publication. doi:10.1080/09669582.2018.1560455

Haites, E. (2009). Linking emissions trading schemes for international aviation and shipping emissions. *Climate Policy, 9*(4), 415–430. doi:10.3763/cpol.2009.0620

Israr, H. A., Yap, C. T. M., Wong, K. J., & Yahya, M. Y. (2016). Compressive properties of Hawaiian Gold Timber Bamboo under different conditions. *Journal of Advanced Research in Applied Mechanics, 25*(1), 10–18.

Jyoti Nath, A., & Das, A. K. (2012). Ecological implications of village bamboo as global climate change mitigation strategy. *International Journal of Climate Change Strategies and Management, 4*(2), 201–215. doi:10.1108/17568691211223169

Koens, K., Postma, A., & Papp, B. (2018). Is overtourism overused? Understanding the impact of tourism in a city context. *Sustainability (Basel), 10*(12), 4384. doi:10.3390/su10124384

Lal, R. (2004). Soil carbon sequestration impacts on global climate change and food security. *Science, 304*(5677), 1623–1627. doi:10.1126/science.1097396 PMID:15192216

Lal, R. (2007). Carbon sequestration. *Philosophical Transactions of the Royal Society of London. Series B, Biological Sciences, 363*(1492), 815–830. doi:10.1098/rstb.2007.2185 PMID:17761468

Lenzen, M., Sun, Y. Y., Faturay, F., Ting, Y. P., Geschke, A., & Malik, A. (2018). The carbon footprint of global tourism. *Nature Climate Change, 8*(6), 522–528. doi:10.1038/s41558-018-0141-x

Lessard, G., & Chouinard, A. (1980). Bamboo research in Asia. In *Proceedings of a workshop held in Singapore,* 28-30 May 1980. International Development Research Centre.

Loehr, J., Becken, S., Nalau, J., & Mackey, B. (2022). Exploring the multiple benefits of ecosystem-based adaptation in tourism for climate risks and destination well-being. *Journal of Hospitality & Tourism Research (Washington, D.C.), 46*(3), 518–543. doi:10.1177/1096348020944438

Lu, C.-W., Huang, J.-C., Chen, C., Shu, M.-H., Hsu, C.-W., & Tapas Bapu, B. R. (2021). An energy-efficient smart city for Sustainable Green Tourism Industry. *Sustainable Energy Technologies and Assessments, 47*, 101494. doi:10.1016/j.seta.2021.101494

Luo, X., Wang, Q., Zheng, B., Lin, L., Chen, B., Zheng, Y., & Xiao, J. (2017). Hydration properties and binding capacities of dietary fibers from bamboo shoot shell and its hypolipidemic effects in mice. *Food and Chemical Toxicology*, *109*(2), 1003–1009. doi:10.1016/j.fct.2017.02.029 PMID:28237776

Mostafanezhad, M., & Evrard, O. (2018). Geopolitical ecologies of tourism and the transboundary haze disaster in Thailand, Laos and myanmar. Community, Environment and Disaster Risk Management, 19. doi:10.1108/S2040-726220180000019003

Pang, S. F., McKercher, B., & Prideaux, B. (2013). Climate change and tourism: An overview. *Asia Pacific Journal of Tourism Research*, *18*(1–2), 4–20. doi:10.1080/10941665.2012.688509

Rasoolimanesh, S. M., Ramakrishna, S., Hall, C. M., Esfandiar, K., & Seyfi, S. (2020). A. *Journal of Sustainable Tourism*, 1–21. doi:10.1080/09669582.2020.1775621

Ross, S., & Wall, G. (1999). Ecotourism: Towards congruence between theory and Practice. *Tourism Management*, *20*(1), 123–132. doi:10.1016/S0261-5177(98)00098-3

Ruiz-Pérez, M., Maoyi, F., Xiaosheng, Y., & Belcher, B. (2001). Bamboo forestry in China: Toward environmentally friendly expansion. *Journal of Forestry*, *99*(7).

Scheyvens, R. (1999). Ecotourism and the empowerment of local communities. *Tourism Management*, *20*(2), 245–249. Advance online publication. doi:10.1016/S0261-5177(98)00069-7

Scott, D., Peeters, P., & Gössling, S. (2010). Can tourism deliver its "aspirational" greenhouse gas emission reduction targets? *Journal of Sustainable Tourism*, *18*(3), 393–408. doi:10.1080/09669581003653542

Scurlock, J. M. O., Dayton, D. C., & Hames, B. (2000). Bamboo: An overlooked biomass resource? *Biomass and Bioenergy*, *19*(4), 229–244. doi:10.1016/S0961-9534(00)00038-6

Seraphin, H., Bah, M., Fyall, A., & Gowreesunkar, V. G. B. (2020). Tourism education in France and sustainable development goal 4 (quality education). *Worldwide Hospitality and Tourism Themes*, *13*(1), 139–147. Advance online publication. doi:10.1108/WHATT-08-2020-0083

Snyman, S. L. (2012). The role of tourism employment in Poverty Reduction and community perceptions of conservation and tourism in Southern Africa. *Journal of Sustainable Tourism*, *20*(3), 395–416. doi:10.1080/09669582.2012.657202

Sundararaju, V. (2021, May 10). *A wonder bamboo from Tamil Nadu to mitigate climate change*. Down To Earth. https://www.downtoearth.org.in/blog/climate-change/a-wonder-bamboo-from-tamil-nadu-to-mitigate-climate-change-76851

Syafiqah Ismail, I., Adilla Rashidi, N., & Yusup, S. (2022). Production and characterization of bamboo-based activated carbon through single-step H 3 PO 4 activation for CO 2 capture. *Environmental Science and Pollution Research International*, *29*, 1–7. doi:10.1007/s11356-021-15030-x/Published PMID:34189693

Turner, C. (2018). *Global issue topics for essays and research.* Academic Press.

Vehbi, B. O., & Doratli, N. (2010a). Assessing the impact of tourism on the physical environment of a small coastal town: Girne, Northern Cyprus. *European Planning Studies*, *18*(9), 1485–1505. doi:10.1080/09654313.2010.492587

Vorosmarty, C. J., Green, P., Salisbury, J., & Lammers, R. B. (2000). Global water resources: Vulnerability from climate change and population growth. *Science*, *289*(5477), 284–288.

Wang, R., Guo, Z., Cai, C., Zhang, J., Bian, F., Sun, S., & Wang, Q. (2021). Practices and roles of bamboo industry development for alleviating poverty in China. *Clean Technologies and Environmental Policy*, *23*(6), 1687–1699. doi:10.1007/s10098-021-02074-3

World Health Organization. (2022, June 16). *World Mental Health Report: Transforming mental health for all.* World Health Organization. https://www.who.int/publications/i/item/9789240049338

Wysocki, W. P., Clark, L. G., Attigala, L., Ruiz-Sanchez, E., & Duvall, M. R. (2015). Evolution of the bamboos (Bambusoideae; Poaceae): A full plastome phylogenomic analysis. *BMC Evolutionary Biology*, *15*(1), 50. Advance online publication. doi:10.1186/s12862-015-0321-5 PMID:25887467

Xu, X., & Reed, M. (2018). Perceived pollution and inbound tourism for Shanghai: A panel VAR approach. *Current Issues in Tourism*, *22*(5), 601–614. doi:10.1080/13683500.2018.1504898

Yang, S. J., Zhang, Y. J., Goldstein, G., Sun, M., Ma, R. Y., & Cao, K. F. (2015). Determinants of water circulation in a woody bamboo species: Afternoon use and night- time recharge of culm water storage. *Tree Physiology*, *35*(9), 964–974. doi:10.1093/treephys/tpv071 PMID:26232783

Yap, C. T. M., & Yahya, M. Y. (2016). Compressive properties of Hawaiian gold timber bamboo under different conditions. *Journal of Advanced Research in Applied Mechanics*, *25*(1), 10–18.

Zea Escamilla, E., Habert, G., & Wohlmuth, E. (2016). When CO2 counts: Sustainability assessment of industrialized bamboo as an alternative for social housing programs in the Philippines. *Building and Environment*, *103*, 44–53. doi:10.1016/j.buildenv.2016.04.003

Zeng, C., Lyu, B., Deng, S., Yu, Y., Li, N., Lin, W., Li, D., & Chen, Q. (2020). Benefits of a three-day bamboo forest therapy session on the physiological responses of university students. *International Journal of Environmental Research and Public Health*, *17*(9), 3238. doi:10.3390/ijerph17093238 PMID:32384727

Zhang, N., Ren, R., Zhang, Q., & Zhang, T. (2020). Air pollution and tourism development: An interplay. *Annals of Tourism Research*, *85*(March), 103032. Advance online publication. doi:10.1016/j.annals.2020.103032

Chapter 11
The Phenomenon of Eco-Anxiety and Distress Related to Climate Change and Environmental Degradation

Santhosh Kumar Rajamani
https://orcid.org/0000-0001-6552-5578
MAEER MIT Pune's MIMER Medical College and Dr. BSTR Hospital, India

Radha Srinivasan Iyer
https://orcid.org/0000-0001-7387-4401
SEC Centre for Independent Living, India

ABSTRACT

Eco-anxiety refers to chronic fear of environmental doom due to climate change and ecological harm. It stems from lived experiences, media coverage, scientific reports, and government inaction. Climate anxiety is rising, especially among youth. Exposure to climate disasters and species/habitat loss can lead to climate grief. Eco-anxiety can manifest as fatalism, avoidance, activism, or compulsions. It reflects conflict between protecting nature and feelings of helplessness. Chronic climate distress heightens risks of depression, PTSD, panic, and substance abuse. Addressing eco-anxiety requires climate education and communication, mental health resources, government climate action, and research on coping strategies. Recognition of eco-anxiety establishes climate mitigation as imperative for collective psychological wellbeing.

DOI: 10.4018/979-8-3693-2177-5.ch011

INTRODUCTION

Climate change poses a significant worldwide concern, exerting diverse impacts on both the environment and human health. Several ramifications of climate change on mental health. The escalation of stress, anxiety, and depression can be attributed to the prevailing uncertainty and apprehension regarding forthcoming circumstances, the loss of livelihoods, residences, and loved ones, as well as exposure to traumatic incidents such as disasters, conflicts, and violence. Survivors of severe weather phenomena such as hurricanes, floods, droughts, heatwaves, and wildfires face an augmented susceptibility to post-traumatic stress disorder (PTSD), substance abuse, and suicide. Elevated temperatures, water scarcity, food insecurity, and social instability contribute to heightened levels of aggression and violence. The disruption of health systems, infrastructure, and communication networks results in diminished accessibility to mental health services and support. The erosion of social and cultural links, loss of identity and sense of belonging, and lower quality of life can lead to a decrease in coping capacity and resilience (Clayton et al., 2017).

UNDERSTANDING ECO-ANXIETY

The term "Eco-anxiety" is employed to delineate the psychological anguish and anxiety that individuals may undergo as a consequence of their apprehensions around the environmental catastrophe and climate change. The interconnectedness of ecological and mental well-being, emphases the importance of recognizing and places eco-anxiety as a valid psychological concern (Clayton et al., 2017).

Eco-anxiety, as delineated by the American Psychiatric Association (APA) as a persistent apprehension regarding environmental cataclysm resulting from the perceived irreversible consequences of climate change and the concomitant apprehension for one's own and future generations' well-being, is a logical reaction to the climate crisis. Climate-related disasters have a significant impact on individuals across many age groups, with a special emphasis on those who are directly affected by such events. Many individuals in the younger generation experience feelings of frustration, anger, and pessimism as a result of observing environmental destruction and perceived insufficient efforts by those in positions of authority. Nevertheless, it is worth noting that eco-anxiety can also function as a productive and adaptive pro-environmental reaction for individuals who redirect their discomfort towards constructive endeavors such as campaigning for policy modifications, endorsing sustainable practices, and engaging in community initiatives (Verplanken, B. et al 2020) .

The phenomenon under consideration is distinguished by emotions of powerlessness, apprehension, or concern regarding the prospects of the globe, the ramifications of climate change on ecosystems, and the welfare of forthcoming generations. The experience of eco-anxiety can vary in intensity, ranging from a moderate sense of concern to a more severe state of discomfort. This psychological phenomenon can exert notable impacts on an individual's mental well-being. This statement underscores the interdependence between environmental concerns and mental health, underscoring the significance of simultaneously tackling the ecological and psychological dimensions of this complex issue.

Eco-anxiety is a prevalent emotional response that arises in individuals due to the environmental issues encountered by our global civilization. Recognizing and affirming one's emotions is of paramount significance, while concurrently seeking constructive mechanisms to manage and proactively address them is equally imperative. By engaging in such actions, individuals have the potential to enhance their psychological well-being while also contributing to personal growth and environmental sustainability.

The experience of residing in an era characterized by climate change exerts a profound influence on individuals' emotional well-being. Observing the dynamic transformations occurring in our natural environment, while experiencing personal culpability or observing apathy towards climate issues and the lack of action from elected authorities, can elicit a range of emotional responses. These may encompass sentiments of anger, irritation, apprehension, helplessness, and despair. The experience can elicit discomfort, a sense of being overwhelmed, and a state of paralysis (Rai, 2023).

Across the globe, there is a notable surge in the prevalence of eco-anxiety among the younger population. Individuals perceive the opportunity to address the pressing global environmental crisis diminishing, yet frequently experience a sense of helplessness in effectuating substantial transformation (Rajamani & Iyer, 2023a). According to recent data, a significant majority of individuals within the younger demographic, specifically over 70%, experience a sense of hopelessness when confronted with the climate crisis. However, it is noteworthy that a much smaller proportion, specifically 26%, possess a perceived understanding of how they may actively contribute to addressing this pressing issue.

The experience of eco-anxiety varies among individuals. Individuals residing in regions directly impacted by climate-related calamities, such as severe flooding, wildfires, and prolonged droughts, may have heightened levels of psychological and emotional anguish. There exists a prevailing sentiment of discontent and disillusionment among Generation Z regarding the perceived lack of proactive measures taken by individuals occupying positions of authority (Norgaard, K. M. 2011).

It is important to acknowledge that eco-anxiety can be considered a sensible response, since it serves as an internal mechanism alerting us to the presence of underlying issues. Recognizing these emotions is crucial; however, redirecting them towards constructive endeavors for the betterment of the environment can serve to mitigate a portion of the anguish. Engaging in activities such as pushing for policy changes, endorsing sustainable practices, or actively participating in community initiatives are all valuable contributions that collectively contribute to the advancement of a more resilient future for our planet (Wake, 2022).

THE PSYCHOLOGICAL IMPACT OF CLIMATE CHANGE

Climate change poses a significant worldwide concern, exerting diverse impacts on both the environment and human health. Several ramifications of climate change on mental health include:

- The escalation of stress, anxiety, and depression can be attributed to the prevailing uncertainty and apprehension regarding forthcoming circumstances, the loss of livelihoods, residences, and loved ones, as well as exposure to traumatic incidents such as disasters, conflicts, and violence.
- Survivors of severe weather phenomena such as hurricanes, floods, droughts, heatwaves, and wildfires face an augmented susceptibility to post-traumatic stress disorder (PTSD), substance abuse, and suicide.
- Elevated temperatures, water scarcity, food insecurity, and social instability contribute to heightened levels of aggression and violence.
- The disruption of health systems, infrastructure, and communication networks results in diminished accessibility to mental health services and support.1. The erosion of social and cultural links, loss of identity and sense of belonging, and lower quality of life can lead to a decrease in coping capacity and resilience.

Climate change and environmental degradation are two urgent concerns that exert a substantial influence on the human psyche. The worldwide concerns at hand exert significant impacts on both the physical environment and the mental health and well-being of individuals.

The impact of climate change and environmental degradation on human cognition is primarily manifested through the occurrence of natural disasters. The increasing frequency and intensity of extreme weather events as a consequence of climate change can lead to traumatic experiences, loss, and displacement for humans. This phenomenon has the potential to give rise to a variety of mental health conditions, including post-traumatic stress disorder, anxiety disorders, and depressive disorders.

Moreover, it is worth noting that environmental degradation might exert a direct influence on mental well-being. The association between air and water pollution and heightened prevalence of respiratory ailments and neurological conditions has been established. This phenomenon can give rise to physiological manifestations that can impact an individual's psychological well-being and general life satisfaction.

In addition, it is important to acknowledge that climate change and environmental degradation can exert indirect influences on mental health, in addition to their direct effects. As the availability of resources diminishes and disputes surrounding them emerge, individuals may encounter heightened levels of stress, anxiety, and potentially even engage in acts of violence. The aforementioned phenomenon has the potential to result in societal and economic disturbances, hence exerting a notable influence on individuals' mental well-being.

The deterioration of the environment might engender feelings of pessimism and despair. As individuals bear witness to the ecological degradation of the earth and perceive their limited agency in mitigating it, they may undergo emotional responses characterized by sadness, wrath, and anxiety pertaining to the prospects of the future. The consideration of the psychological impact that climate change and environmental degradation can impose on underprivileged people is of significant importance. Low-income and minority populations frequently experience a disproportionate impact from these challenges, potentially encountering supplementary stressors such as limited resource availability and instances of prejudice.

In light of these issues, it is imperative for both individuals and communities to proactively address the consequences of climate change and environmental degradation through mitigation and adaptation measures. This may encompass pushing for policies aimed at addressing these concerns, mitigating individual carbon footprints, and endorsing sustainable practices.

Furthermore, it is imperative for mental health professionals to possess knowledge regarding the ramifications of climate change and environmental degradation on their patients, and to integrate this understanding into their therapeutic interventions. This encompasses the provision of assistance to persons who are affected by trauma resulting from natural disasters, the mitigation of mental health consequences arising from pollution, and the promotion of resilience in the context of environmental adversities.

Climate change poses a significant risk not only to our physical health, but also to our mental well-being. The recognition of the psychological ramifications of climate change and the implementation of measures to address its causes and adapt to its consequences are of utmost significance. Mutual support and assistance can be sought among individuals. Collectively, we possess the capacity to confront this formidable obstacle and forge a more salubrious and enduring prospect for both humanity and our environment.

In summary, the impacts of climate change and environmental deterioration on the human psyche are substantial. Recognizing and addressing the implications of environmental factors on mental well-being is of paramount importance for individuals, communities, and mental health professionals alike, as it serves to improve both environmental and mental well-being (Berry et al., 2018).

UNDERSTANDING THE ROOT CAUSES

The root causes of eco-anxiety, including the fear of a bleak environmental future, the loss of biodiversity, and the overwhelming uncertainty about climate impacts. It highlights how these factors contribute to the development and intensification of climate-related distress. Numerous studies have demonstrated that the occurrence of natural disasters and the manifestation of extreme weather events, both attributable to climate change, can exert a substantial influence on individuals' mental well-being. The occurrence of these events has the potential to induce trauma, anxiety, sadness, and post-traumatic stress disorder (PTSD) in persons who undergo them (Rajamani & Iyer, 2023a).

Figure 1. Eco-anxiety an original illustration created by prompting "draw a picture of a person with anxiety related to environmental degradation and drastic climate changes" Stable diffusion generative image AI engine available at https://stablediffusionweb.com/. Images generated by the model are explicitly licensed in public domain.

An investigation conducted after to the occurrence of Hurricane Katrina revealed that 49% of the respondents exhibited indications of depression, while 25% exhibited indications of post-traumatic stress disorder (PTSD). In a similar vein, a research investigation conducted on individuals impacted by the floods in Thailand in 2011 revealed that approximately 27% of the participants exhibited symptoms indicative of post-traumatic stress disorder (PTSD), while approximately 30% reported experiencing symptoms associated with depression ("Past Climate Changes and Human Adaptation," 2019).

Figure 2. Swelling global carbon dioxide emissions, showing increased greenhouse gas emissions, leading to heat trapping and increasing temperatures worldwide. Original illustration of authors based on dataset from DataHub. "CO_2 PPM - Trends in Atmospheric Carbon Dioxide" dataset, retrieved on 10-Dec-23.

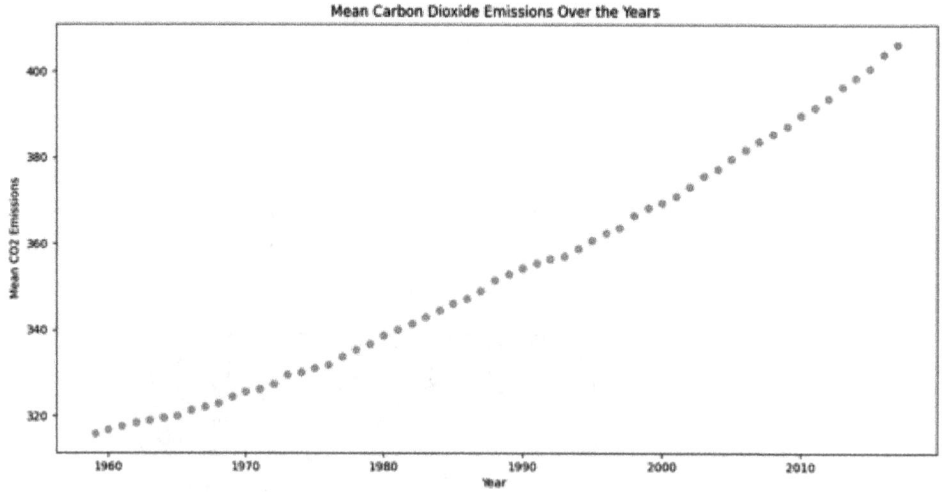

In conjunction with the immediate consequences of climate catastrophes, the persistent risk of impending calamities can also engender psychological distress. The experience of living in locations prone to disasters can induce considerable stress and anxiety in individuals due to the unpredictability and apprehension surrounding the timing and location of future calamities (Brulle, R. J. et al, 2012).

In addition, the mental health consequences of climate catastrophes can be enduring, as they involve the displacement and disruption of homes, livelihoods, and relationships. This assertion holds particular validity when considering underprivileged communities, as they frequently experience a disproportionate impact from such occurrences.

In order to mitigate the mental health consequences arising from climate catastrophes, it is imperative to offer assistance and allocate resources to individuals impacted by these events. This encompasses provisions for mental health services, community support programs, and disaster preparedness and response plans that prioritize mental health. Furthermore, by effectively tackling the underlying factors contributing to climate change and implementing measures to minimize its consequences, it is possible to avert or diminish the occurrence and intensity of climate-related disasters, hence mitigating their influence on mental well-being (Hannart & Naveau, 2018).

CLIMATE COMMUNICATION AND MEDIA INFLUENCE

Communication plays a pivotal role in shaping public perception and, consequently, eco-anxiety. This section investigates the role of media in disseminating information and fostering awareness about climate change while discussing the responsibility of media outlets in framing climate narratives responsibly.

PSYCHOLOGICAL RESILIENCE AND COPING STRATEGIES

The chapter provides readers with a toolbox of coping strategies to manage eco-anxiety and climate distress. It explores the potential of eco-therapy, mindfulness practices, and emotional resilience-building techniques as valuable tools for individuals seeking to navigate the emotional challenges of the climate crisis.

CASE STUDIES IN ECOANXIETY AND DISTRESS DUE TO ENVIRONMENTAL DEGRADATION

The adverse consequences of climate change have significant social and economic implications, often intensifying pre-existing disparities, disputes, and infringements on human rights. The phenomenon of climate change has the potential to significantly impact the accessibility and cost-effectiveness of fundamental necessities, including but not limited to sustenance, water resources, energy provisions, and healthcare services. Additionally, it has the potential to heighten the likelihood of acts of violence, forced displacement, migratory patterns, and the occurrence of people trafficking (World Economic Forum 2022).

In the year 2020, the World Food Program (WFP) issued a warning indicating that a considerable number of individuals, over 270 million, faced a significant threat of experiencing acute hunger. This perilous situation arose as a result of the compounding impacts of various factors, including conflict and other related circumstances (Kostopoulou & Giannakopoulos, 2023).

The decline in biodiversity and ecosystems results in a decrease in the ability of nature to adapt and function effectively, hence impacting its services. Numerous species are confronted with the imminent threat of extinction as a result of factors such as habitat degradation, overexploitation, pollution, the introduction of exotic species, and the adverse impacts of climate change. In the year 2021, the International Union for Conservation of Nature (IUCN) made revisions to its Red List of Threatened Species, resulting in the inclusion of an additional 8,000 species. Consequently, the cumulative count of species classified as threatened now exceeds 38,000 (Rajamani & Iyer, 2023a).

The escalating occurrence and severity of natural calamities, including floods, wildfires, droughts, hurricanes, and heatwaves. These occurrences have the potential to result in fatalities, forced migration, physical destruction, and psychological distress for a significant number of individuals and animal populations. Moreover, they have the potential to impede food security, water supply, health systems, and infrastructure. In the year 2023, the Philippines experienced the impact of three formidable typhoons within a duration of three weeks, resulting in a significant loss of life exceeding 300 individuals and affecting millions of people (Raz, 2022).

The escalating sea levels and coastal erosion pose significant risks to the sustenance and cultural heritage of numerous inhabitants residing in low-lying islands and coastal regions. Certain communities are currently encountering the imminent threat of displacement from their residences and territories due to the advancing presence of water bodies. In the year 2022, the Maldives proclaimed a state of emergency due to the inundation of over 80 islands caused by elevated tidal levels (Cazenave & Remy, 2011).

The Ecoanxiety due Frequent Floods in North-Eastern Provinces of India

On the 4th of October 2023, an intense precipitation event resulted in the overflow of the South Lhonak lake, a glacial lake situated in the state of Sikkim, located in the northeastern region of India. The South Lhonak Lake is a body of water that is formed by the accumulation of meltwater from the Lhonak glacier, which is blocked by a moraine. The flash flood was ascribed to the combination of heavy rainfall and the melting of the Lhonak glacier. The death toll resulting from flash floods in India has increased following the rupture of the glacier lake attributed to increasing global temperatures, an event that had been previously cautioned by experts over a span of several years.

This event subsequently led to a glacial lake outburst flood in the Teesta River.At midnight, the floodwaters arrived at the Teesta III Dam located near Chungthang, causing the dam to be destroyed within a matter of minutes, prior to the commencement of gate opening procedures. The rise in water levels downstream in the River Teesta resulted in significant damage, with an increase of up to twenty feet observed. The recorded number of fatalities amounted to approximately 75 individuals, while the hydroelectric Teesta III Dam and 15 river bridges were subsequently destroyed as a result of the flooding.

According to a prior research conducted by the Indian Institute of Technology (IIT) Guwahati, there has been an observed escalation in the occurrence of flash floods in Sikkim, India, which can be attributed to the effects of climate change. The research additionally revealed a notable augmentation of 14% in the magnitude of precipitation inside the area during recent decades.

The report proposed that it is imperative for the government to implement appropriate strategies in order to address the impacts of climate change and effectively decrease the production of greenhouse gases. Additionally, it was suggested that everyone residing in the area should be duly informed about the potential hazards linked to flash floods and be provided with appropriate training to effectively address such occurrences.

The Ecoanxiety due to Frequent Landslides in India

Landslides are geological phenomena characterized by the downward movement of substantial quantities of soil, rock, or other materials along a slope. In mountainous locations such as India, avalanches have the potential to inflict substantial harm on both life and property. India ranks among the top five countries globally that are susceptible to landslides. This vulnerability is shown in the annual occurrence of at least one fatality per 100 square kilometres as a result of landslide incidents.

Landslides can be initiated by various sources, including but not limited to intense precipitation, seismic events, and anthropogenic actions such as deforestation and mining. Minimizing the risk of landslides necessitates the implementation of preventive measures, including the construction of retaining walls, the strategic planting of trees, and the avoidance of construction activities on steep slopes.

Several causes contribute to the occurrence of landslides in India, including variations in rainfall patterns, seismic events, slope instability, deforestation, mining activities, construction practices, and human interventions. The user's text is not clear and does not provide any information. The Northwest and Northeast parts of India, specifically the Himalayas, along with the Western Ghats, are characterized by high susceptibility to landslides, which occur frequently and exhibit significant severity(Yonetani & Gordon, 2001).

Landslides are a frequently occurring natural calamity in the Indian subcontinent, particularly in the context of the monsoon period. The Himalayan region has heightened susceptibility to landslides as a consequence of its rugged topography and substantial precipitation. In the past few years, the region has experienced a series of landslides, resulting in substantial human casualties and property damage (Afzali et al., 2020).

Immense Landslides in India in Last 50 Years

The following is a brief account of immense landslides that happened in India in last half-century:

- The Guwahati landslide in Assam in 1948 resulted in a significant loss of life, with more than 500 individuals perishing and an entire hamlet being engulfed as a consequence of intense precipitation.
- The Darjeeling landslide in West Bengal in 1968 resulted in a significant loss of life, with thousands of individuals perishing. Additionally, the event caused severe damage to infrastructure, including the fragmentation of a 60 km roadway into 91 separate sections, primarily owing to flooding.
- The Malpa landslide in Uttarakhand in 1998 resulted in the loss of over 380 lives and the complete destruction of the village of Malpa. This catastrophic event was caused by a series of landslides that were prompted by intense rainfall.
- The Mumbai landslide that occurred in Maharashtra in the year 2000 resulted in the unfortunate loss of around 67 lives. Additionally, the local railway services were severely impacted as a consequence of ground erosion caused by high rainfall[1].
- The Amboori landslide in Kerala, which occurred in 2001, resulted in the unfortunate loss of approximately 40 lives. This event stands as the most severe landslide recorded in the history of Kerala, mostly attributed to the substantial precipitation experienced during that period[1].
- The Kedarnath landslide in Uttarakhand in 2013 resulted in a significant loss of life, with an estimated death toll exceeding 5700 individuals. Additionally, the floods and subsequent landslides had a profound impact on over 4200 villages in the vicinity of the Kedarnath temple.
- The Malin landslide in Maharashtra in 2014 resulted in the loss of about 151 lives, with an additional 100 individuals reported as missing. This tragic event was triggered by a landslide that transpired in a village near Malin, mostly attributed to the occurrence of intense precipitation.

- In the year 2021, a significant landslide event transpired inside the Raigad area of Maharashtra, India, resulting in the unfortunate loss of more than 200 lives. at the year 2020, a landslide occurrence took place at Munnar, Kerala, India, resulting in the unfortunate loss of life of a minimum of 70 individuals.
- The Mumbai landslide incident in 2021 resulted in a tragic loss of life, with a minimum of 32 individuals perishing and numerous others sustaining injuries. The victims became trapped beneath residential structures that succumbed to landslides triggered by intense precipitation in the Chembur and Vikhroli areas, which are suburban regions situated inside Mumbai.
- The Kinnaur landslide in 2021 resulted in the unfortunate loss of 13 lives and the entrapment of other individuals beneath the debris. This calamitous event occurred in a Himalayan area located in the northern region of India. A series of descending rock masses dislodged and forcefully displaced a truck, a bus, and several other motor vehicles along a segment of the highway located in the state of Himachal Pradesh.
- The floods and landslides that occurred in Kerala in 2020 resulted in a significant loss of life, with over 125 individuals losing their lives. Additionally, the adverse weather conditions, characterized by severe rainfall, led to the displacement of over 300,000 individuals throughout the state. The districts that saw the most severe impact were Idukki, Wayanad, Malappuram, and Kozhikode.
- The Malin landslide of 2019 resulted in the loss of about 151 lives, with an additional 100 individuals reported as missing. This tragic event was triggered by a landslide that took place in a village in Malin, which was attributed to intense precipitation. The landslide event in Maharashtra's history was considered to be among the most severe.

Landslides pose a significant risk to the prosperity and security of India. Hence, the implementation of efficient strategies pertaining to landslide hazard zonation, mitigation, management, monitoring, and early warning systems holds significant importance. The issuance of guidelines pertaining to these factors by the National Disaster Management Authority (NDMA) has been accompanied by the allocation of funding for diverse research and mapping projects on landslides conducted by various organizations (Pihkala, 2020a).

The Ecoanxiety Due to Frequent Floods in Pakistan

The phenomenon of climate anxiety has had a profound impact on the communities in Pakistan that have been hit by floods, particularly those who have experienced the severe repercussions of these exceptional natural disasters. In the previous year, a significant number of individuals, namely over 33 million, experienced direct effects, and a substantial proportion, amounting to 20.6 million, were in need of immediate humanitarian aid. The consequences of the floods are still recent, as an estimated, 1.5 million individuals continue to be displaced. Although certain immediate needs such as food, housing, and water have been partially met, the issue of climate fear inside these communities has received less attention in media coverage (Rummukainen, 2012).

Despite making a relatively little contribution to the global climate catastrophe, Pakistan is still very susceptible to the adverse impacts of climate change. The escalating frequency of flood occurrences has ensnared many communities inside an interminable cycle of forced relocation and profound despondency. The mental well-being of individuals is adversely affected by the exhaustion resulting from climate-induced calamities. In a state of apprehension regarding an ambiguous future, individuals are confronted with the stark realization that they possess insufficient measures in place to confront a subsequent flood event. Homes that possess a delicate structural integrity may potentially become uninhabitable, so leaving their occupants without well-defined strategies for securing alternative refuge (Doherty & Clayton, 2016).

Gul Khatoon, an individual who has experienced the adverse effects of flooding, conveyed her apprehensions on the future that her offspring will inherit, considering the persistent devastation inflicted upon our village by flood events. The individual in question delivered a child within an improvised shelter located alongside a road during the floods of 2022, thereby drawing attention to the extremely challenging circumstances experienced by other individuals.

The current state of affairs in Pakistan highlights the imperative for international intervention in order to tackle the issue of climate change and its disproportionate consequences on marginalized communities. In the pursuit of resilience and adaptability, it is imperative to recognize and confront the issue of eco-anxiety, as it plays a pivotal role in constructing a sustainable future (Hayes et al., 2018).

The Ecoanxiety and Phenomenon of Farmer Suicides in India

The agricultural sector in India encounters various obstacles, including erratic weather patterns, inadequate irrigation facilities, and market volatility, all of which can lead to severe hardships for numerous farmers. The occurrence of these suicides frequently arises from a confluence of variables, encompassing financial hardship, agricultural yield decline, escalating indebtedness, inadequate credit availability, and restricted support networks (Sharma et al., 2021).

The phenomenon of farmer suicides in India is a distressing and intricate matter characterized by a multitude of causes and ramifications. The causes of suicides has been attributed to numerous reasons, namely:

- Climate change-induced environmental degradation has been identified as a significant factor impacting crop yields, water availability, and pest infestation. The agricultural sector faces considerable financial burdens attributed to the high costs associated with inputs such as seeds, fertilizers, pesticides, equipment, and labour (Tiwari et al., 2021).
- The agricultural industry is further challenged by the low and volatile prices of agricultural produce, resulting in reduced income and profitability for farmers. Farmers, particularly those who are small-scale or marginalized, encounter difficulties in accessing adequate and timely credit, insurance, and social security (Sharma et al., 2021).
- The burden of debt and interest from private moneylenders and banks often subjects farmers to harassment and humiliation. The implementation of government policies and schemes, including minimum support prices, crop insurance, loan waivers, and compensation for crop losses, has been characterized by poor execution.
- Personal and familial issues, such as illness, marital conflict, alcoholism, and domestic violence, also contribute to the challenges faced by farmers. The prevalence of mental health conditions, including stress, anxiety, despair, and feelings of hopelessness, has been widely documented in academic literature (Mishra & Nagaraj, 2016).

The web search results also indicate that instances of farmer suicides have been documented in other states within India, with Maharashtra experiencing the highest prevalence. According to the National Crime Records Bureau (NCRB) data, the year 2019 witnessed a significant number of suicides among farmers and agricultural laborers, surpassing 10,000 cases. This demographic constituted approximately 7.4% of the overall suicide victims in India. According to the statistics provided by the National Crime Records Bureau (NCRB), there was an observed 18% increase in the incidence of suicides among agricultural laborers in the year 2020 as compared to the preceding year of 2019 (National Crime Records Bureau, 2020). Conversely, a marginal decrease in the number of suicides among farmers and cultivators was also noted.

The issue of farmer suicides in India is a significant concern of both social and economic nature, demanding immediate attention and concerted efforts from all relevant parties involved. Addressing the underlying factors contributing to agricultural distress and ensuring sufficient assistance and relief measures are provided to the farming community are of paramount significance. It is imperative to foster consciousness and compassion regarding the predicament faced by farmers and their families, as well as to advocate for the enhancement of their mental health and overall well-being (Burke et al., 2015).

Various measures have been undertaken to tackle this situation, encompassing governmental initiatives, programs aimed at alleviating debt burdens, and the provision of mental health assistance to farmers. Nevertheless, it is imperative to acknowledge that the matter at hand is intricate and encompasses various aspects, necessitating the implementation of comprehensive measures . These measures should encompass enhancements in agricultural methodologies, the establishment of equitable pricing mechanisms for agricultural produce, and the provision of enhanced social and psychological support for individuals engaged in farming activities (Mishra, 2014).

The Ecoanxiety and Frequent Floods in South India

South India has long been acquainted with the rhythm of monsoon rains, a source of life and sustenance for millions. However, in recent years, this rhythm has turned tumultuous, with increasingly frequent and intense floods causing widespread devastation. This environmental crisis has not only displaced communities and destroyed livelihoods but also triggered a surge in a new phenomenon. In South India, the frequency and severity of floods have been linked to climate change, with rising sea levels and erratic rainfall patterns exacerbating the problem (Kumar et al., 2022). Witnessing the destruction of homes, crops, and ecosystems firsthand has triggered eco-anxiety amongst individuals and communities, particularly those directly affected by the floods (Hinds & Sparks, 2017).

The 2021 floods in South India encompass a sequence of flood events linked to depression and a low pressure system, resulting in extensive disruptions over the Indian states of Tamil Nadu, Andhra Pradesh, and the neighboring Sri Lanka. The precipitation event commenced on the first of November in the year 2021 throughout the region of Tamil Nadu(Baudon, 2021). The occurrence of floods can be attributed to the occurrence of exceptionally intense precipitation events. This calamity resulted in the unfortunate loss of life, with a minimum of 41 individuals reported as casualties in both India and Sri Lanka (Parmar et al., 2022).

One study conducted in Kerala, a state heavily impacted by floods, found a significant correlation between flood exposure and eco-anxiety symptoms .The study reported feelings of fear, hopelessness, and guilt among individuals who had experienced the floods, highlighting the emotional toll of such events (Kumar, K. K. et. al. 2022)

MITIGATING ECO-ANXIETY

Addressing eco-anxiety necessitates a multifaceted approach, encompassing individual and community-level interventions. Mindfulness practices, cognitive reframing, and building resilience can empower individuals to manage their anxiety and cope with the emotional burden. Fostering social support networks, promoting environmental activism, and facilitating dialogue about climate change can further strengthen communities and create a sense of collective agency.

However, individual and community-level efforts alone are insufficient. Addressing the root cause of eco-anxiety requires systemic changes. Implementing effective climate change mitigation and adaptation strategies, investing in sustainable development, and prioritizing environmental protection are crucial for reducing environmental threats and fostering hope for the future(Santhosh Kumar Rajamani & Radha Srinivasan Iyer, 2023b).

Additionally, integrating eco-anxiety awareness into public health initiatives and ensuring accessible mental health services are essential for providing support to individuals and communities grappling with the emotional impact of a changing climate.

The escalating climate crisis necessitates a holistic approach that addresses not only the physical consequences but also the mental health challenges it poses. By acknowledging eco-anxiety, promoting individual and community resilience, and implementing comprehensive climate action plans, we can pave the way for a future where environmental responsibility and emotional well-being go hand in hand. This future requires a collective effort, one that calls for collaboration between individuals, communities, policymakers, and institutions to ensure a sustainable and mentally healthy world for generations to come.

INDIVIDUAL ACTION

Eco-anxiety is a natural reaction to the environmental challenges that we face as a global society There is a growing recognition of the need to address eco-anxiety alongside the physical and economic challenges posed by climate change. This requires a multi-pronged approach that includes:

- Providing mental health support: Psychological services and community-based initiatives can help individuals cope with eco-anxiety and build resilience.
- Promoting adaptation strategies: Supporting communities in adopting climate-resilient infrastructure and livelihoods can foster a sense of empowerment and control(Santhosh Kumar Rajamani & Radha Srinivasan Iyer, 2023c).
- Raising awareness: Educating the public about eco-anxiety and its symptoms is crucial. This can help to normalize the experience and reduce feelings of isolation and shame.

- Addressing the root cause: Taking action to mitigate climate change is crucial to prevent future floods and reduce the psychological burden on communities (Kaushik, 2022)

It is important to acknowledge and validate an individual's feelings, but also to find healthy ways to cope and take action. To improve mental health, but also make a positive difference for a individual human being and the planet (Ardenne, 2021).

COMMUNITY AND COLLECTIVE ACTION

This section highlights how communities worldwide are uniting to address eco-anxiety collectively. It underscores the importance of support networks, local initiatives, and grassroots movements in fostering resilience and hope (Ardenne, 2021). There are several strategies for managing eco-anxiety and mitigating its adverse impacts on mental well-being. Several strategies can be identified, including:

- It is advisable to actively pursue precise and dependable information regarding climate change and its potential remedies. However, it is important to exercise caution and refrain from consuming sensationalized or inflated media reports, since they have the potential to heighten anxiety levels.
- Building community resilience: Fostering social connections, promoting environmental action, and empowering communities to take charge of their own adaptation efforts can help to build resilience and reduce anxiety.
- It is advisable to engage in proactive measures aimed at diminishing one's individual carbon footprint and actively support environmental initiatives that are in line with one's personal values and objectives. This can contribute to an enhanced sense of empowerment and optimism towards future prospects(Rehling, 2022).
- Foster connections with those who possess similar worries and provide mutual assistance. Participating in a community group, an online forum, or a social movement can engender a feeling of affiliation and unity among individuals.
- It is advisable to consult with a qualified expert if one's eco-anxiety begins to impede their everyday functioning or results in notable suffering. Therapeutic professionals has the ability to facilitate the examination of one's emotions, confront pessimistic cognitions, and cultivate adaptive mechanisms for managing distress.
- Addressing climate change: Ultimately, the most effective way to reduce eco-anxiety is to address the root cause of the problem: climate change. This

requires global cooperation and a commitment to transitioning to a low-carbon economy.

- Engage in self-care practices and participate in activities that elicit feelings of joy and relaxation. This practice has the potential to facilitate emotional equilibrium and replenish one's energy levels. Several activities that have been suggested for promoting well-being include meditation, yoga, engaging in physical activity, pursuing hobbies, and immersing oneself in natural environments (Clayton, S. et. al. 2017).

While the future may seem uncertain, there is reason for hope. The growing awareness of eco-anxiety and the increasing focus on mental health are positive steps in the right direction. By taking action to address both the physical and emotional impacts of climate change, we can build a more resilient future for all communities facing the challenges of a changing climate (Pihkala, 2020b)

CONCLUSION NAVIGATING THE WATERS OF ECO-ANXIETY

Eco-anxiety, a relatively new phenomenon, presents a significant mental health challenge exacerbated by the urgency of climate change. Recognizing the complex emotions triggered by environmental threats, such as fear, despair, anger, and anxiety, is crucial for developing effective coping mechanisms and building resilience.

While acknowledging the gravity of the situation, focusing solely on the negative aspects of climate change can perpetuate feelings of hopelessness and paralysis. Therefore, striking a balance between acknowledging the reality of the crisis and fostering a sense of agency and hope through individual and collective action is critical.

Addressing eco-anxiety requires a multi-pronged approach. On an individual level, mindfulness practices, cognitive reframing, and building resilience can help manage anxiety and promote coping. At the community level, fostering social support networks, promoting environmental activism, and facilitating dialogue about climate change can empower individuals and create a sense of collective agency.

Policy and advocacy also play a vital role in addressing eco-anxiety. Implementing policies that mitigate climate change, promote sustainable development, and invest in climate adaptation strategies are crucial for reducing environmental threats and fostering hope for the future. Additionally, advocating for mental health services and integrating eco-anxiety awareness into public health initiatives can provide crucial support to individuals and communities dealing with the emotional impact of climate change.

By recognizing the reality of eco-anxiety and taking concrete steps at individual, community, and societal levels, we can navigate the turbulent waters of climate change with greater awareness, resilience, and hope. Remember, we are not powerless in the face of this crisis. By working together, we can build a more sustainable and resilient future for all.

REFERENCES

Afzali, R., GharehBeygi, M., & Yazdanpanah Dero, Q. (2020). Climate changes and food policies: Economic pathology. *Climate Risk Management*, *30*, 100249. doi:10.1016/j.crm.2020.100249

Ardenne, P. (2021, June 4). From Eco-anxiety to Resilience. *Critique D'art*, *56*(56), 47–59. doi:10.4000/critiquedart.76145

Baudon, P. (2021, May). A scoping review of interventions for the treatment of eco-anxiety. In Analysis, 5(1), 82–85. doi:10.1016/j.inan.2021.02.005

Berry, H. L., Waite, T. D., Dear, K. B., Capon, A. G., & Murray, V. (2018). Human health and climate change: Leverage points for adaptation in urban environments. *International Journal of Environmental Research and Public Health*, *15*(6), 1232. PMID:29891795

Brulle, R. J., Carmichael, J., & Jenkins, J. C. (2012). Shifting public opinion on climate change: An empirical assessment of factors influencing concern over climate change in the US, 2002–2010. *Climatic Change*, *114*(2), 653–672. doi:10.1007/s10584-012-0403-y

Burke, M. B., Miguel, E., Satyanath, S., Dykema, J. A., & Lobell, D. B. (2015). Climate change in the Fertile Crescent and implications of the recent Syrian drought. *Proceedings of the National Academy of Sciences of the United States of America*, *112*(11), 3241–3246. doi:10.1073/pnas.1421533112 PMID:25733898

Cazenave, A., & Remy, F. (2011, August 17). Sea level and climate: Measurements and causes of changes. *Wiley Interdisciplinary Reviews: Climate Change*, *2*(5), 647–662. doi:10.1002/wcc.139

Clayton, S., Manning, C., Krygsman, K., & Speiser, M. (2017). *Mental health and our changing climate: impacts, implications, and guidance*. American Psychological Association.

Clayton, S., Manning, C. M., Krygsman, K., & Speiser, M. (2021). *Mental health and our changing climate: Impacts, implications, and guidance*. American Psychological Association. doi:10.1037/e507892021-001

DataHub. (n.d.). *CO2 PPM - Trends in Atmospheric Carbon Dioxide* [Dataset]. https://datahub.io/core/co2-ppm

Doherty, T. J., & Clayton, S. (2016). Climate change and mental health: Causal pathways, policy implications, and research priorities. *EcoHealth*, *13*(4), 654–657.

Hannart, A., & Naveau, P. (2018, July). Probabilities of Causation of Climate Changes. *Journal of Climate*, *31*(14), 5507–5524. doi:10.1175/JCLI-D-17-0304.1

Hayes, K., Blashki, G., Wiseman, J., Burke, S., & Reifels, L. (2018). Climate change and mental health: Risks, impacts and priority actions. *International Journal of Mental Health Systems*, *12*(1), 1–12. doi:10.1186/s13033-018-0210-6 PMID:29881451

Hinds, J., & Sparks, P. (2017). Mental health impacts of floods. *Current Psychiatry Reports*, *19*(11), 1–10.

Kaushik, A. (2022, October 28). Eco anxiety. *British Dental Journal*, *233*(8), 588–588. doi:10.1038/s41415-022-5164-z PMID:36307681

Kostopoulou, E., & Giannakopoulos, C. (2023, February 21). Projected Changes in Extreme Wet and Dry Conditions in Greece. *Climate (Basel)*, *11*(3), 49. doi:10.3390/cli11030049

Kumar, K. K., Prasanna, V. L., & Roy, P. S. (2022). Floods in South India: An overview of causes and consequences. In *Climate Change Impacts and Challenges in India* (pp. 151–165). Springer.

Mishra, N., & Nagaraj, K. (2016). Suicides of farmers in Maharashtra, India: A qualitative exploration of their causes. *Journal of Injury & Violence Research*, *8*(1).

Mishra, S. (2014). Agrarian distress and farmers' suicides in India: Is there any way out? *Indian Journal of Agricultural Economics*, *69*(3), 375–388.

National Crime Records Bureau. (2020). *Accidental deaths & suicides in India (ADSI) report-2019 & ADSI report-2020 (provisional data)*. Ministry of Home Affairs.

Norgaard, K. M. (2011). *Living in denial: Climate change, emotions, and the denial of grief*. Oxford University Press. doi:10.7551/mitpress/9780262015448.001.0001

Parmar, S., Kumar, M., & Kumar, A. (2022). The impact of Kerala floods on mental health: A cross-sectional study. *International Journal of Disaster Risk Reduction*, *70*, 102732.

Past climate changes and human adaptation. (2019, November). *Past Global Changes Magazine, 27*(2). doi:10.22498/pages.27.2.83

Pihkala, P. (2020a, September 23). Anxiety and the Ecological Crisis: An Analysis of Eco-Anxiety and Climate Anxiety. *Sustainability (Basel), 12*(19), 7836. doi:10.3390/su12197836

Pihkala, P. (2020b, December 4). Eco-Anxiety and Environmental Education. *Sustainability (Basel), 12*(23), 10149. doi:10.3390/su122310149

Rai, A. (2023, August). Understanding Eco-Anxiety and Eco-Grief: A Comprehensive Review. *International Journal of Research Publication and Reviews, 4*(8), 681–686. doi:10.55248/gengpi.4.823.50353

Rajamani, S. K., & Iyer, R. S. (2023b). Networks in healthcare: A systematic review. *BioMedInformatics, 3*(2), 391–404. doi:10.3390/biomedinformatics3020026

Raz, S. (2022, July 4). Ethics of Climate Responsibility and Eco-Anxiety. *Psychoanalytic Dialogues, 32*(4), 348–350. doi:10.1080/10481885.2022.2090814

Rehling, J. T. (2022, October 17). Conceptualising eco-anxiety using an existential framework. *South African Journal of Psychology. Suid-Afrikaanse Tydskrif vir Sielkunde, 52*(4), 472–485. doi:10.1177/00812463221130898

Rummukainen, M. (2012, February 16). Changes in climate and weather extremes in the 21st century. *Wiley Interdisciplinary Reviews: Climate Change, 3*(2), 115–129. doi:10.1002/wcc.160

Santhosh Kumar Rajamani & Radha Srinivasan Iyer. (2023a). Declining Freshwater Species Biodiversity. In N. K. Chourasia & K. Chahal (Eds.), *Perspectives on Global Biodiversity Scenarios and Environmental Services in the 21st Century* (pp. 22–56). IGI Global. doi:10.4018/978-1-6684-9034-1.ch002

Santhosh Kumar Rajamani & Radha Srinivasan Iyer. (2023c). Use of Python Modules in Ecological Research. In P. O. de Pablos, M. N. Almunawar, & M. Anshari (Eds.), *Perspectives on the Transition Toward Green and Climate Neutral Economies in Asia* (pp. 182–206). IGI Global. doi:10.4018/978-1-6684-8613-9.ch011

Sharma, D., Singh, S., Singh, B., Singh, A. V. J., Singh, R., Singh, R., Singh, G., Singh, J., Singh, B., Singh, R., & ... (2021). Farmer suicides in India: A policy-induced disaster of epic proportions. *Economic and Political Weekly, 56*(2), 13–16.

Tiwari, R. K., Kumar, A., & Joshi, P. K. (2021). Farmer suicides in India: A comprehensive analysis of official data (1995–2019). *World Development Perspectives, 23*, 100345.

Verplanken, B., Marks, E., & Dobromir, A. I. (2020). On the nature of eco-anxiety: How constructive or unconstructive is habitual worry about global warming? *Journal of Environmental Psychology*, *72*, 101528. doi:10.1016/j.jenvp.2020.101528

Wake, B. (2022, September 29). Understanding eco-anxiety. *Nature Climate Change*, *12*(10), 884–884. doi:10.1038/s41558-022-01497-0

World Economic Forum. (2022). *Global risks report 2023*. Retrieved April 27, 2023, from https://www.weforum.org/publications/global-risks-report-2023/in-full/1-global-risks-2023-today-s-crisis

World Food Programme. (2020). *WFP chief warns of hunger pandemic as COVID-19 spreads* (Statement to UN Security Council). Retrieved April 27, 2023, from https://www.wfp.org/news/wfp-chief-warns-hunger-pandemic-covid-19-spreads-statement-un-security-council

Yonetani, T., & Gordon, H. B. (2001, February 8). Abrupt changes as indicators of decadal climate variability. *Climate Dynamics*, *17*(4), 249–258. doi:10.1007/PL00007926

Chapter 12
Ethical Issues for Climate Change and Mental Health

Pankaj Bhambri
https://orcid.org/0000-0003-4437-4103
Guru Nanak Dev Engineering College, Ludhiana, India

Sita Rani
https://orcid.org/0000-0003-2778-0214
Guru Nanak Dev Engineering College, Ludhiana, India

ABSTRACT

This chapter delves into the complex ethical dimensions surrounding the intersection of climate change and mental well-being. As global environmental shifts increasingly impact the psychological health of individuals and communities, ethical considerations become paramount in guiding effective and culturally sensitive interventions. The chapter navigates the complexities of integrating Indigenous perspectives, fostering global collaboration, and balancing immediate mental health needs with long-term sustainability goals. Case studies and ethical dilemmas offer real-world insights into the challenges faced by practitioners, highlighting the need for nuanced decision-making processes. This chapter aims to enhance the resilience, fairness, and ethical foundation of addressing the significant effects of climate change on mental health and well-being by analyzing and tackling these ethical considerations.

1. INTRODUCTION

The perception of climate change has evolved beyond being just an environmental issue, as its significant effects on mental health have garnered increasing global attention (Adams, M. 2021). The complex correlation between climate change and

DOI: 10.4018/979-8-3693-2177-5.ch012

mental health is diverse, involving both immediate and indirect repercussions that greatly impact individuals and society. The increasing occurrence and severity of climate-related phenomena, such as hurricanes, wildfires, and droughts, directly contribute to mental health consequences, including trauma, post-traumatic stress disorder (PTSD), and anxiety disorders. These sudden catastrophes can result in the displacement of people, the death of loved ones, and the ruin of homes, creating a complicated interaction of emotional and psychological difficulties for those impacted (Davidson, J.M., & Mill, R.W. 2020).

In addition to these immediate impacts, the gradual, pervasive changes associated with climate change pose indirect threats to mental well-being. The looming uncertainty about the future, exacerbated by phenomena like rising sea levels and temperature extremes, contributes to a growing sense of eco-anxiety—a chronic fear and worry about the environmental consequences. This emotional burden is not confined to individuals directly impacted by climate events; it permeates societies globally, leading to a collective psychological toll. Low-income communities and disadvantaged groups, which are vulnerable populations, are frequently impacted to a greater extent. This exacerbates pre-existing gaps in mental health and highlights the ethical obligation to address climate justice within the realm of mental health (IPCC 2018).

Moreover, the intricate web of climate change impacts extends beyond immediate trauma, influencing social structures, economic stability, and community dynamics. Changes in climate patterns can disrupt livelihoods, strain social support systems, and escalate resource conflicts, creating conditions conducive to chronic stress and mental health disorders (Kaur, A. et al. 2012). The importance of fully understanding the relationship between climate change and mental health is emphasized by the wider impact it has on society. This understanding should go beyond individual psychological reactions and also take into account the complex connections within communities and the larger social structure. Understanding this intricate interaction is crucial for formulating ethical and efficient approaches to address and manage the mental health issues stemming from climate change.

As scientists delve deeper into the various effects of climate change, it becomes more and more clear that addressing mental health in the context of a changing climate is not only a scientific necessity but also a moral obligation. The correlation between climate change and mental health highlights the interdependence of worldwide issues and the pressing need to implement ethical frameworks that direct comprehensive remedies (Pachauri, R.K., & Reisinger, A. 2007). In this context, mental health professionals, policymakers, and communities must collaboratively navigate the ethical dimensions of climate change adaptation and mitigation, striving for solutions that prioritize the well-being of individuals and societies in the face of a rapidly changing world.

1.1 Ethical Considerations in Addressing Climate Change Impacts on Mental Well-Being

As the consequences of climate change manifest globally, the ensuing environmental alterations, extreme weather events, and resource scarcities significantly impact individuals' mental health. Ethical considerations in addressing these impacts extend beyond immediate psychological support to encompass broader questions related to justice, equity, and responsibility (Reser, J.P., & Swim, J.K. 2011). This backdrop necessitates a comprehensive exploration of the ethical dimensions associated with climate change and mental well-being interventions.

- Disproportionate Vulnerability: An important ethical concern arises from the unequal susceptibility of specific populations to the psychological consequences of climate change. Marginalized communities, which are frequently already burdened by socioeconomic injustices, experience a greater weight of responsibility when confronted with environmental pressures. The ethical imperative lies in recognizing and addressing these disparities, ensuring that mental health interventions are not only accessible but also culturally sensitive. Questions of distributive justice arise, urging a closer examination of how resources, support systems, and mental health services are allocated to diverse communities affected by climate change.
- Interconnectedness of Ecosystems and Mental Health: Ethical considerations also arise from the acknowledgment of the interconnectedness of ecosystems and mental well-being. Climate change disrupts ecosystems, altering landscapes, and diminishing the availability of natural resources. As individuals' connections to their environments are compromised, ethical dilemmas emerge regarding the responsibilities of societies, governments, and industries in contributing to environmental degradation. Recognizing this interdependence underscores the importance of adopting a holistic and ecologically-informed ethical framework that considers the reciprocal relationship between environmental health and mental well-being.
- Psychological Impacts of Climate Grief and Eco-Anxiety: The emotional burden of climate bereavement and eco-anxiety adds an additional level of ethical intricacy. People are struggling with intense emotions caused by the decline in biodiversity, environmental deterioration, and uncertainty on what lies next. Mental health professionals must navigate ethical considerations in validating and addressing these emotions, ensuring that interventions are not only therapeutic but also grounded in an understanding of the existential challenges posed by climate change. Acknowledging the ethical duty to

alleviate suffering, professionals must develop strategies that empower individuals to cope with the emotional toll of climate-related distress.

- Global Responsibility and Inter-Generational Equity: Finally, ethical considerations extend globally, emphasizing the shared responsibility to address climate change impacts on mental health. This involves not only mitigating current challenges but also ensuring inter-generational equity. Ethical frameworks must grapple with the long-term consequences of environmental degradation and the implications for the mental well-being of future generations. The question of how to balance present needs with future sustainability poses ethical challenges that demand collaborative, informed decision-making on a global scale. Addressing the mental health impacts of climate change necessitates a commitment to ethical principles that transcend geographical boundaries, emphasizing the collective responsibility to safeguard the well-being of current and future populations.

2. CLIMATE CHANGE AND VULNERABLE POPULATIONS

Climate change intensifies pre-existing vulnerabilities and disproportionately affects underprivileged groups globally. Marginalized people, such as low-income communities, indigenous groups, and those with limited access to resources, frequently have increased vulnerabilities and difficulties due to environmental changes. The displacement of vulnerable communities is exacerbated by rising sea levels, extreme weather events, and changing ecosystems, resulting in heightened pressures on mental health (Singh, S. et al. 2013). Moreover, these populations may lack the financial means and infrastructure to adapt or recover swiftly, intensifying the psychological toll.

Addressing the intersection of climate change and mental health for vulnerable populations requires a nuanced understanding of socio-economic factors, cultural contexts, and historical injustices (Weber, E.U. 2010). The ethical considerations are profound, as mental health disparities are intertwined with broader social inequalities. Mental health professionals must navigate the ethical imperative to provide culturally sensitive and equitable support to communities that bear a disproportionate burden of climate change impacts. Additionally, ethical frameworks should guide interventions that empower vulnerable populations, recognizing their agency and resilience. Collaborative efforts between mental health professionals, community leaders, and policymakers are essential to develop adaptive strategies that not only alleviate the mental health consequences of climate change but also promote social justice and environmental equity for the most vulnerable.

2.1 Climate Change Disproportionately Affects the Vulnerable Populations

Climate change exacerbates existing vulnerabilities and socio-economic disparities, disproportionately impacting marginalized and vulnerable populations around the world. Many marginalized communities, such as low-income persons, indigenous groups, and those residing in vulnerable ecosystems, frequently face a dearth of resources and infrastructure necessary to effectively respond to the impacts of climate change. They might inhabit regions susceptible to severe weather phenomena, such as floods, storms, or droughts, which are exacerbated by climate change (Leiserowitz, A. et al. 2019). Limited access to education, healthcare, and economic opportunities further compounds their vulnerability, making it challenging for these populations to recover from climate-related disruptions.

Secondly, vulnerable populations are often reliant on climate-sensitive sectors such as agriculture, fisheries, and forestry for their livelihoods. Changes in precipitation patterns, temperature extremes, and the increasing frequency of natural disasters directly threaten their food security and economic stability. Indigenous communities, in particular, face the loss of traditional practices and cultural heritage due to altered ecosystems (Gupta, S. et al. 2007). Additionally, the unequal distribution of environmental hazards and exposure to pollution tends to concentrate in low-income neighborhoods, placing vulnerable populations at higher risk for climate-related health issues. As a result, the impact of climate change on these communities extends beyond environmental concerns to include economic, social, and health dimensions, perpetuating and deepening existing inequalities. Addressing climate change effectively requires recognizing and prioritizing the needs of these vulnerable populations to ensure a just and equitable response.

2.2 Ethical Considerations in Providing Mental Health Support to Marginalized Communities

Providing mental health support to marginalized communities is a crucial aspect of addressing health disparities and promoting social justice. However, this endeavor comes with a set of ethical considerations that mental health professionals must navigate to ensure equitable and effective care. Here are key ethical considerations in this context:

- Cultural Competence and Sensitivity: Mental health providers must possess cultural competence and demonstrate sensitivity towards the distinct experiences, values, and customs of marginalized communities. Ethical practice requires an understanding of how cultural factors influence mental health perceptions, help-seeking behaviors, and the expression of distress.

- Access to Services: Ensuring equitable access to mental health services is an ethical imperative. Barriers such as economic disparities, language barriers, and geographic location must be actively addressed. Mental health professionals should advocate for policies and practices that reduce systemic barriers to accessing care for marginalized populations.
- Power Dynamics and Empowerment: Recognizing and addressing power imbalances between mental health professionals and marginalized individuals is crucial. Ethical practice involves promoting empowerment and fostering collaboration. Mental health interventions should aim to amplify the voices of marginalized communities, allowing them to actively participate in decision-making regarding their mental health care.
- Intersectionality: Acknowledging and understanding the intersectionality of identities within marginalized communities is an ethical imperative. Individuals may face multiple forms of discrimination, and mental health support should be inclusive and sensitive to these intersecting identities. Mental health professionals should strive to provide intersectional and culturally competent care that considers the complexities of an individual's identity.
- Trauma-Informed Care: Many individuals in marginalized communities may have experienced systemic trauma. Mental health professionals must adopt trauma-informed approaches that prioritize safety, trust, and collaboration. Ethical considerations include avoiding re-traumatization, respecting autonomy, and incorporating trauma-sensitive interventions into mental health support.
- Advocacy for Social Justice: Ethical mental health practice extends beyond individual therapy to encompass advocacy for social justice. Mental health professionals should actively engage in efforts to address systemic inequalities and discriminatory practices. Advocating for policies that promote social equity, challenging stigma, and participating in community initiatives align with the ethical responsibility of mental health professionals.
- Informed Consent and Collaboration: Informed consent is crucial, and mental health professionals must communicate information in a manner that is culturally and linguistically appropriate. Collaborative decision-making ensures that individuals from marginalized communities actively participate in their treatment planning, fostering a sense of agency and self-determination.
- Confidentiality and Trust: Maintaining confidentiality is essential, but mental health professionals must also be transparent about any limitations due to legal or safety considerations. Building and maintaining trust is foundational

to ethical mental health practice, and this is especially critical when working with communities that may have historical reasons for mistrust.

3. MENTAL HEALTH PROFESSIONALS' RESPONSIBILITIES

Mental health professionals bear the ethical responsibility of staying informed about the evolving field of climate psychology, recognizing the unique vulnerabilities of diverse populations, and tailoring therapeutic interventions to navigate climate-related stressors (Kjellstrom, T. et al. 2009). This include the promotion of resilience, coping strategies, and community support systems. In addition, mental health professionals have a vital role in advocating for policies that prioritize mental well-being in response to climate change. They also engage in interdisciplinary collaborations to improve the effectiveness of interventions. Their responsibilities extend beyond clinical practice to contribute to the broader discourse on sustainable practices, equitable resource distribution, and ethical considerations within the intersection of climate change and mental health.

3.1 Ethical Obligations of Mental Health Professionals in the Face of Climate-Related Mental Health Challenges

As mental health practitioners address the intricate issues of mental health problems associated to climate change, it is crucial to thoroughly analyze their ethical responsibilities (Prerna et al. 2012). First and foremost, it is imperative to be knowledgeable and up-to-date on the developing comprehension of the psychological consequences of climate change. This necessitates ongoing education and professional development to equip practitioners with the knowledge and skills required to address the unique stressors emerging from environmental shifts.

Furthermore, mental health professionals bear the responsibility of acknowledging and dismantling potential biases within the therapeutic relationship that may arise in the context of climate-related concerns. It becomes imperative to recognize and address any pre-existing prejudices or misconceptions that could impact the quality of care provided to individuals grappling with climate-induced anxieties, grief, or trauma.

The ethical considerations extend to the development of culturally sensitive and contextually relevant interventions. Mental health professionals must be attuned to the diverse ways in which different communities experience and express climate-related distress. Cultivating cultural competence ensures that therapeutic approaches are respectful, inclusive, and effective, fostering a deeper connection between the professional and the client.

Furthermore, a dedication to social justice emerges as a fundamental ethical principle. Psychiatrists and psychologists are expected to actively support policies and structural modifications that target the underlying factors of climate change and its psychological consequences, especially in populations that are disproportionately impacted by environmental emergencies (Jamieson, D. 2014). This advocacy goes beyond the therapeutic session, motivating professionals to participate in wider activities aimed at reducing the global impact of climate change on mental well-being.

Mental health professionals have a range of ethical responsibilities when it comes to addressing mental health difficulties associated to climate change. These responsibilities include ongoing education and self-examination, as well as practicing in a culturally sensitive and socially equitable manner. As climate change increasingly shapes the psychological landscape, mental health professionals play a pivotal role in fostering resilience, healing, and sustainable coping strategies for individuals and communities confronting the mental health repercussions of environmental shifts.

3.2 Roles of Psychotherapy, Counseling, and Community Interventions

In addressing the mental health implications of climate change, mental health professionals must carefully consider the diverse therapeutic modalities and community interventions that can effectively support individuals and communities facing environmental stressors. Psychotherapy and counseling play pivotal roles in providing individuals with a safe space to express and process climate-related anxieties, grief, and trauma (Green, B., & Johnson, M. 2020). Therapeutic interventions should be tailored to the unique needs of clients, recognizing the spectrum of emotional responses to environmental challenges.

- Psychotherapy: Psychotherapy offers a structured and confidential setting where individuals can explore the emotional and psychological impact of climate change on their lives. Therapists may employ various therapeutic modalities, such as mindfulness-based approaches, or ecotherapy, cognitive-behavioral therapy (CBT) depending on the client's preferences and therapeutic goals. Integrating eco-psychological principles into therapy can foster a deeper connection between individuals and the natural world, facilitating healing and resilience.
- Counseling: Counseling interventions extend beyond individual therapy sessions to include family and group counseling. Climate change can strain familial relationships and community dynamics, making family and group counseling essential for addressing shared concerns and fostering collective resilience. Mental health professionals may collaborate with community

organizations to provide counseling services that address the unique challenges faced by families and groups affected by environmental changes.

- Community Interventions: Recognizing the interconnectedness of mental health and community well-being, mental health professionals should actively engage in community-level interventions. This involves collaborating with local organizations, schools, and grassroots initiatives to develop and implement programs that enhance community resilience and coping mechanisms. Community interventions may include workshops, support groups, and educational initiatives that empower individuals to navigate the psychological impact of climate change collectively.

Moreover, mental health professionals can contribute to the development of community-based disaster response plans that prioritize mental health and emotional well-being. This proactive approach ensures that communities are better equipped to address the psychological fallout of climate-related events, fostering a sense of preparedness and cohesion.

4. CLIMATE CHANGE MITIGATION AND MENTAL HEALTH

As the global community grapples with the escalating impacts of climate change, the intersection of climate change mitigation efforts and mental health outcomes becomes increasingly evident (Bhambri, P., & Nischal, P. 2013). Climate change mitigation, which involves reducing or preventing the emission of greenhouse gases to limit global temperature rise, has far-reaching implications for mental health. Understanding and addressing the mental health aspects of mitigation strategies are crucial for the overall well-being of individuals and communities.

- Psychological Impact of Mitigation Policies: The implementation of climate change mitigation policies, while essential for long-term environmental sustainability, can evoke a range of psychological responses in individuals and communities. These responses may include anxiety, fear, or even a sense of loss as people adapt to changes in lifestyle, industries, and energy sources. Mental health professionals play a pivotal role in helping individuals navigate these emotions and build resilience in the face of transformative societal shifts.
- Eco-Anxiety and Climate Grief: Climate change mitigation efforts can contribute to the rise of eco-anxiety and climate grief. Eco-anxiety arises from the existential threat posed by climate change, while climate grief involves mourning the loss of ecosystems, biodiversity, and familiar landscapes.

Mental health professionals are tasked with recognizing and addressing these emerging psychological challenges, offering therapeutic support to individuals experiencing distress related to environmental changes and the necessary but often disruptive mitigation measures.

- Community-Level Mental Health Considerations: At the community level, the implementation of mitigation strategies can impact mental health through changes in employment patterns, economic structures, and local ecosystems. Mental health professionals can contribute to community-based interventions that address these broader impacts, ensuring that the mental health implications of climate change mitigation are considered in the planning and implementation of policies. Community engagement becomes integral in fostering a sense of collective agency and well-being amidst change.
- Communication and Public Perception: Effective communication about climate change mitigation is crucial to shaping public perception and reducing anxiety associated with uncertainties (Cohen, G.L. 2018). Mental health professionals can collaborate with environmental scientists and policymakers to develop communication strategies that convey the urgency of mitigation efforts while providing coping mechanisms for individuals and communities. This involves fostering a sense of empowerment through informed decision-making and emphasizing the potential mental health co-benefits of sustainable practices.
- Social Equity and Justice: An ethical dimension of climate change mitigation involves ensuring social equity and justice. Mental health professionals advocate for policies that consider the disparate impacts of mitigation measures on vulnerable populations, addressing issues of accessibility, affordability, and inclusivity in mental health support services. This advocacy aligns with a broader commitment to fostering mental well-being within a framework of environmental and social justice.

5. CLIMATE GRIEF AND ECO-ANXIETY

Climate change mitigation efforts, while primarily focused on reducing greenhouse gas emissions and adapting to environmental shifts, inherently possess profound implications for mental health (Bhambri, P., & Gupta, O.P. 2013). The pursuit of sustainable practices and policies aimed at curbing climate change has the potential to positively impact mental well-being on a global scale. As communities engage in collective initiatives to reduce their carbon footprint and transition towards environmentally responsible practices, individuals may experience a heightened sense of purpose, community connection, and hope for the future. However, the psychological toll of climate change mitigation should not be underestimated.

Adjusting to new environmental norms, confronting the realities of a changing world, and addressing potential economic and social disruptions can induce stress, anxiety, and grief. Mental health professionals play a crucial role in supporting individuals and communities through this transition, advocating for policies that prioritize both ecological sustainability and psychological resilience, thereby fostering a harmonious balance between environmental stewardship and mental health.

5.1 Ethical Aspects of Addressing Climate Grief and Eco-Anxiety

The growing prevalence of climate grief and eco-anxiety necessitates a thoughtful exploration of the ethical considerations surrounding mental health professionals' engagement with these profound emotional experiences. Acknowledging and validating these feelings becomes a crucial ethical imperative, as individuals grapple with the emotional toll of environmental degradation and the apprehension about an uncertain future (Edwards, S., & Harris, M. 2017). Mental health professionals must tread ethically when addressing climate grief, ensuring a compassionate and empathetic approach that validates the unique and sometimes overwhelming emotions associated with witnessing the ecological crisis unfold. Respect for diverse coping mechanisms and the creation of a safe space for individuals to express their grief ethically form the foundation of therapeutic interventions.

Simultaneously, the emergence of eco-anxiety, characterized by intense worry and fear related to ecological concerns, prompts ethical considerations regarding the boundaries of mental health practice. Practitioners must delicately balance raising awareness about the environmental crisis while mitigating the potential for overwhelming anxiety. Ethical guidelines dictate the need for psychoeducation that empowers individuals to channel their eco-anxiety into constructive actions, fostering a sense of agency in the face of environmental challenges (Nisbet, M.C. 2018). Mental health professionals are ethically obligated to address the root causes of eco-anxiety, advocating for systemic change, and promoting sustainable practices to alleviate the emotional burdens associated with ecological concerns.

In navigating the ethical landscape of climate grief and eco-anxiety, mental health professionals play a pivotal role in fostering resilience, coping strategies, and a sense of community support. This ethical commitment extends beyond the therapeutic setting to advocacy for societal changes that contribute to a more sustainable and ecologically responsible future, aligning mental health interventions with a broader ethical imperative to address the interconnected challenges of climate change and mental well-being.

5.2 Role of Mental Health Professionals in Validating and Addressing the Emotions

In the face of escalating climate grief and eco-anxiety, mental health professionals play a crucial role in validating and addressing the complex emotional responses that individuals may experience (Goyal, F. et al. 2012). One key aspect of their role involves creating a compassionate and judgment-free space where individuals can openly express their emotions related to climate change. Validation becomes an ethical imperative as mental health professionals acknowledge the legitimacy of feelings such as sadness, despair, or fear, recognizing them as valid and rational responses to the unprecedented ecological challenges confronting our planet. By validating these emotions, mental health professionals help reduce the stigma associated with climate-related distress, fostering an environment conducive to healing and emotional resilience.

Addressing these emotions ethically requires mental health professionals to employ a diverse range of therapeutic techniques tailored to the unique needs of each individual (Fisher, E. 2016). From traditional psychotherapy methods to ecotherapy, mindfulness practices, and group support, the goal is to empower individuals to navigate their emotional responses effectively. Ethical considerations demand a nuanced approach, where professionals collaborate with clients to develop coping strategies that are both psychologically beneficial and environmentally sustainable. This involves a delicate balance between acknowledging the severity of the ecological crisis and instilling hope by fostering a sense of agency and engagement in climate action.

Furthermore, mental health professionals have an ethical responsibility to engage in ongoing education and self-reflection to enhance their cultural competence in addressing diverse emotional responses to climate change (Paika, V., & Bhambri, P. 2013). Understanding the intersectionality of emotions within the context of culture, socioeconomic status, and individual lived experiences ensures that therapeutic interventions are inclusive, respectful, and responsive to the unique needs of each client. By validating and ethically addressing the emotions associated with climate grief and eco-anxiety, mental health professionals contribute significantly to the development of resilient individuals and communities facing the psychological impact of environmental challenges.

6. INDIGENOUS PERSPECTIVES AND ETHICS

Indigenous perspectives offer invaluable insights into the ethical dimensions of climate change and mental health, emphasizing the interconnectedness of environmental well-being and holistic health. Rooted in ancestral wisdom and deep connections to the land, Indigenous ethics highlight the importance of sustainable practices and respect for nature. Mental health professionals engaging with Indigenous communities must approach their work with cultural humility, recognizing and valuing diverse worldviews. Ethical considerations involve prioritizing Indigenous voices, respecting traditional healing practices, and incorporating community-driven interventions that align with cultural values (Solomon, S. et al. 2009). Upholding Indigenous rights, land sovereignty, and acknowledging the historical trauma tied to environmental disruptions is an ethical imperative. By weaving Indigenous perspectives into the fabric of mental health interventions, professionals contribute ethically to a more holistic and inclusive approach that honors the wisdom embedded in Indigenous cultures and fosters resilience in the face of climate-induced challenges.

6.1 Recognition of Indigenous Knowledge and Perspectives on Climate Change

The acknowledgment and integration of Indigenous knowledge and perspectives on climate change are essential steps in fostering a more comprehensive and ethical understanding of the complex interplay between environmental shifts and mental well-being. Indigenous communities possess rich, time-tested ecological knowledge that is intricately tied to their cultural and spiritual identities. Recognizing and respecting this wisdom not only acknowledges the historical custodianship of the land but also provides a vital foundation for sustainable mental health practices. Ethically, mental health professionals should engage in collaborative partnerships with Indigenous communities, actively seeking their guidance in addressing the psychological impacts of climate change. This involves valuing Indigenous ways of knowing, respecting traditional ecological practices, and acknowledging the profound implications of environmental disruptions on community health. In weaving Indigenous perspectives into the discourse on climate change and mental health, professionals contribute to a more ethical, culturally sensitive approach that embraces the profound insights embedded in Indigenous knowledge systems and facilitates community-led resilience strategies in the face of environmental challenges.

6.2 Ethical Considerations in Incorporating Indigenous Wisdom Into Mental Health Strategies

Ethical considerations in integrating Indigenous wisdom into mental health strategies are paramount, requiring mental health professionals to approach this collaboration with humility, respect, and a deep understanding of the historical context. Central to this ethical framework is the recognition that Indigenous knowledge systems are not merely tools for addressing mental health challenges but represent a holistic way of life deeply entwined with cultural, spiritual, and environmental contexts. Mental health professionals must engage in meaningful consultation and collaboration with Indigenous communities, seeking explicit consent and respecting their right to control how their knowledge is shared and utilized (van der Linden, S. L. et al. 2015). This involves dismantling colonial power dynamics that have historically marginalized Indigenous voices, ensuring that mental health strategies are co-created and led by the communities themselves.

Furthermore, ethical considerations extend to fostering cultural competence among mental health professionals engaging with Indigenous wisdom. This involves ongoing education and self-reflection to cultivate an awareness of diverse Indigenous cultures, worldviews, and healing practices. Professionals must be attuned to the potential for cultural appropriation and navigate the delicate balance between incorporating Indigenous wisdom into mental health strategies and avoiding the commodification of sacred knowledge. Additionally, ethical practice requires mental health professionals to advocate for policies that uphold Indigenous rights, land sovereignty, and environmental justice, recognizing that the well-being of Indigenous communities is intricately linked to the health of their traditional lands and ecosystems. In embracing these ethical considerations, mental health strategies can authentically incorporate Indigenous wisdom, contributing to a more just, inclusive, and effective approach to addressing the mental health impacts of climate change.

7. INTERNATIONAL COLLABORATION AND ETHICAL STANDARDS

International collaboration in addressing the intersection of climate change and mental health requires a commitment to ethical standards that prioritize inclusivity, cultural sensitivity, and shared responsibility. Mental health professionals engaged in global efforts must adhere to ethical guidelines that transcend geographic boundaries, recognizing the universality of mental well-being and the interconnectedness of environmental challenges. Ethical considerations include ensuring equitable access to mental health resources, respecting diverse cultural contexts, and avoiding the imposition of Western-centric models on non-Western communities. Collaboration should be founded on mutual respect, reciprocity, and the recognition of the unique strengths that each culture brings to the table (Hansen, J., & Sato, M. 2012). Upholding international ethical standards involves advocating for policies that address the root causes of climate change on a global scale and promote mental health resilience across diverse communities, fostering a shared commitment to the well-being of the planet and its inhabitants.

7.1 Global Cooperation in Addressing Climate Change and Mental Health

The interconnected nature of climate change impacts demands international collaboration to develop comprehensive strategies that account for the varied experiences and vulnerabilities of communities worldwide. The World Health Organization (WHO), in partnership with other institutions, has a crucial influence on defining the worldwide discussion on climate change and mental health. The Lancet Countdown on Health and Climate Change and similar initiatives underscore the pressing need for collaborative efforts to comprehensively comprehend and tackle the psychological ramifications of environmental changes.

International organizations, research institutions, and mental health professionals engage in collaborative research to identify patterns, trends, and best practices for mitigating the psychological impacts of climate change. Shared databases and information-sharing platforms facilitate the dissemination of knowledge, fostering a global community of practice dedicated to understanding and addressing climate-related mental health challenges. Moreover, global cooperation involves the exchange of expertise and resources, enabling the implementation of culturally sensitive interventions that respect the unique circumstances of diverse populations.

Ethical considerations in global cooperation include recognizing and respecting the sovereignty of nations and Indigenous communities, ensuring that mental health interventions are inclusive and adapted to local contexts (Kaur, P., & Bhambri, P. 2013). Moreover, international collaboration should prioritize the voices of those most affected, emphasizing participatory approaches that empower communities to contribute to the development of solutions. Global cooperation aims to cultivate a collective comprehension of the ethical aspects arising from the convergence of climate change and mental health. Its objective is to enhance resilience, develop mental well-being, and establish a sustainable future for everyone.

7.2 Ethical Standards for International Collaboration and Resource Allocation

In the realm of international collaboration addressing climate change and mental health, ethical standards are essential to ensure fairness and inclusivity. These standards emphasize equitable resource allocation, acknowledging the varying capacities of nations and communities to address the mental health impacts of environmental changes. Collaborative efforts must prioritize vulnerable populations, recognizing the disproportionate effects of climate change on marginalized communities (Baker, L., & Smith, R. 2019). Ethical considerations involve fostering partnerships that respect cultural diversity, acknowledging the expertise of local practitioners, and avoiding paternalistic approaches. Transparency in resource allocation and decision-making processes is paramount, promoting accountability and trust among collaborating entities. Ultimately, these ethical standards guide international efforts to address climate-related mental health challenges with sensitivity, equity, and a commitment to the well-being of all communities globally.

8. CASE STUDIES AND ETHICAL DILEMMAS

The incorporation of case studies and ethical problems in the discussion on climate change and mental health functions as a crucial pedagogical instrument, providing practical perspectives on the intricacies of tackling psychological consequences in many circumstances (Miller, G. 2017). Mental health professionals can enhance their comprehension of the ethical implications involved in climate-related interventions by conducting thorough examinations of particular cases (U.S. Global Change Research Program. 2018). These case studies examine various situations, ranging from the consequences of severe weather occurrences to the execution of mental health initiatives in culturally vibrant yet ecologically susceptible populations. Ethical dilemmas, woven into these narratives, challenge practitioners to navigate issues such as cultural sensitivity, informed consent, and power dynamics within interdisciplinary collaborations. By engaging with these ethical quandaries, mental health professionals gain valuable perspectives on how to ethically balance intervention effectiveness with cultural relevance, fostering a more nuanced and inclusive approach to addressing climate-related mental health challenges globally.

8.1 Real-Life Case Studies Highlighting Ethical Dilemmas in Climate Change and Mental Health

- Disaster Response and Cultural Sensitivity: In the aftermath of a climate-related disaster, mental health professionals are deployed to provide support in a culturally diverse community. Ethical dilemmas arise as practitioners grapple with the challenge of tailoring interventions to respect cultural norms, belief systems, and traditional healing practices. Striking a balance between immediate psychological needs and cultural sensitivity becomes crucial, highlighting the ethical imperative of honoring diverse perspectives in the face of urgent mental health crises.
- Community-Led Interventions and Power Dynamics: A mental health initiative is implemented in an Indigenous community affected by environmental changes. Ethical dilemmas emerge as practitioners navigate power dynamics,

ensuring the active participation and leadership of community members in decision-making processes. Balancing the integration of traditional healing practices with modern interventions while avoiding cultural appropriation presents ethical challenges, emphasizing the need for a collaborative and community-driven approach.

- Global Research and Informed Consent: A international research endeavor seeks to examine the psychological consequences of climate change on various populations. Ethical dilemmas surface as researchers navigate informed consent procedures, language barriers, and power differentials between researchers and participants. Ensuring participants' autonomy, understanding, and willingness to engage in research without imposing Western-centric perspectives become central ethical considerations, illustrating the complexities of conducting ethically sound global mental health studies.
- Resource Allocation and Equity: A global mental health initiative allocates resources for climate change adaptation and mental health support. Ethical dilemmas emerge as practitioners grapple with the equitable distribution of resources among nations with varying capacities. Addressing the unique needs of vulnerable populations while avoiding the reinforcement of existing inequalities becomes a significant ethical challenge, highlighting the importance of prioritizing justice and inclusivity in resource allocation decisions.

8.2 Analysis of Decision-Making Processes in Challenging Scenarios

Within the context of climate change and mental health, examining decision-making processes in difficult situations offers a detailed comprehension of the ethical complexities encountered by mental health practitioners. Navigating these scenarios requires a delicate balance between immediate intervention needs and the respect for diverse cultural, social, and environmental contexts. Practitioners often encounter the ethical imperative of swift action in response to mental health crises resulting from climate-related events, all while respecting the autonomy, beliefs, and traditions of the affected communities.

Decision-making processes involve a careful consideration of power dynamics, acknowledging the potential for unintended consequences and the reinforcement of existing disparities. The analysis delves into how practitioners grapple with ethical dilemmas related to resource allocation, ensuring equitable distribution, and prioritizing marginalized populations in the aftermath of environmental disasters. Additionally, it explores the ethical challenges associated with advocating for systemic change while respecting the agency of communities in deciding their own paths to resilience.

Through the examination of decision-making processes in challenging scenarios, mental health professionals gain insights into ethical leadership, cultural humility, and the imperative to foster collaborative, community-driven approaches. This research provides a comprehensive framework for professionals aiming to negotiate the intricate relationship between climate change and mental health. It emphasizes the importance of being sensitive, inclusive, and making ethical decisions.

9. CONCLUSION

When dealing with the complex intersection of climate change and mental health, it is important to address fundamental ethical factors that might provide guidance for mental health providers, governments, and communities. The imperative to prioritize cultural sensitivity, community engagement, and equity underscores the ethical foundations required in addressing the psychological impacts of environmental shifts. The recognition and integration of diverse perspectives, especially those of Indigenous communities, emphasize the importance of honoring traditional wisdom in mental health strategies.

As mental health professionals, it is incumbent upon us to uphold ethical standards that transcend geographic boundaries, fostering global collaboration and inclusivity. This involves continuous education, cultural humility, and an unwavering commitment to justice. Policymakers are called upon to craft frameworks that prioritize both environmental sustainability and mental well-being, ensuring that climate change mitigation strategies are not only effective but also ethically sound.

Communities, as active participants in their own resilience, are encouraged to engage in open dialogue, advocate for their unique needs, and actively participate in decision-making processes. The call to action is a collective one, urging mental health professionals, policymakers, and communities to forge alliances that transcend disciplinary boundaries, acknowledge the ethical nuances of the climate and mental health intersection, and work collaboratively towards a future where mental well-being and environmental stewardship intertwine harmoniously. By adopting these fundamental ethical principles and heeding this urgent plea for action, we engage on a collective endeavor towards a stronger, more inclusive, and morally principled approach to tackling the significant difficulties at the intersection of climate change and mental health.

REFERENCES

Adams, M. (2021). *Climate Change and Mental Health: A Comprehensive Analysis*. Academic Press.

Baker, L., & Smith, R. (Eds.). (2019). *Ethical Considerations in Environmental Psychology*. Routledge.

Bhambri, P., & Gupta, O. P. (2013). A Survey of HPC Applications Appropriate for Execution on DSM and Grid System. *PIMT Journal of Research*, *5*(2), 83–87.

Bhambri, P., & Nischal, P. (2013). *Emergence of the Knowledge Economy and Knowledge Management and their Importance in the Talent Acquisition Front*. Paper presented at the Seminar on NAAC Sponsored National Seminar on Quality Culture in Education.

Cohen, G. L. (2018). *Climate Change Ethics: Environmental Justice and Global Citizenship*. Rowman & Littlefield.

Davidson, J. M., & Mill, R. W. (2020). *Mental Health in a Changing Climate: Impacts, Vulnerabilities, and Adaptation*. Springer.

Edwards, S., & Harris, M. (Eds.). (2017). *Ethical Challenges in Climate Change Communication: Evolving Perspectives in Science, Policy, and Media*. Springer.

Fisher, E. (2016). *Climate Change, Ethics and Human Security*. Cambridge University Press.

Goyal, F., Bhambri, P., & Bansal, P. (2012). Accuracy of Phylogenetic Trees using Distance based Methods. *International Conference on Sports Biomechanics, Emerging Technologies and Quality Assurance in Technical Education*, 400-403.

Green, B., & Johnson, M. (2020). *The Ethics of Climate Change: An Introduction*. Routledge.

Gupta, S., Nischal, P., & Bhmabri, P. (2007). Data Encryption Standard (DES) Algorithm with Diffie-Hellman Key Exchange. *National Conference on Emerging Trends in Communication & IT,* 135-143.

Hansen, J., & Sato, M. (2012). Assessing "Dangerous Climate Change": Required Reduction of Carbon Emissions to Protect Young People, Future Generations and Nature. *PLoS One*, *7*(12), e51068. PMID:23251423

IPCC. (2018). *Global Warming of 1.5°C. An IPCC Special Report on the impacts of global warming of 1.5°C above pre-industrial levels and related global greenhouse gas emission pathways, in the context of strengthening the global response to the threat of climate change, sustainable development, and efforts to eradicate poverty.* Retrieved from https://www.ipcc.ch/sr15/

Jamieson, D. (2014). *Reason in a Dark Time: Why the Struggle to Stop Climate Change Failed—And What It Means for Our Future*. Oxford University Press. doi:10.1093/acprof:oso/9780199337668.001.0001

Kaur, A., Bhambri, P., & Gupta, O. P. (2012). A Novel Technique for Robust Image Segmentation. *International Journal of Advances in Engineering and Technology*, *3*(4), 110–114.

Kaur, P., & Bhambri, P. (2013). Review of Zero Watermarking Algorithms for Text Documents. *PCTE Journal of Computer Sciences*, *10*(1), 77–82.

Kjellstrom, T., Kovats, R. S., Lloyd, S. J., Holt, T., & Tol, R. S. J. (2009). The Direct Impact of Climate Change on Regional Labor Productivity. *Archives of Environmental & Occupational Health*, *64*(4), 217–227. doi:10.1080/19338240903352776 PMID:20007118

Leiserowitz, A., Maibach, E., Rosenthal, S., Kotcher, J., Bergquist, P., Ballew, M., Goldberg, M., & Gustafson, A. (2019). *Climate Change in the American Mind: April 2019*. Yale University and George Mason University. Retrieved from https://climatecommunication.yale.edu/publications/climate-change-in-the-american-mind-april-2019/

Miller, G. (2017). Climate Change and Mental Health: A Scoping Review. *International Journal of Mental Health Systems*, *11*, 43.

Nisbet, M. C. (2018). Communicating Climate Change: Why Frames Matter for Public Engagement. *Environment*, *60*(4), 12–23.

Pachauri, R. K., & Reisinger, A. (Eds.). (2007). *Climate Change 2007: Synthesis Report. Contribution of Working Groups I, II, and III to the Fourth Assessment Report of the Intergovernmental Panel on Climate Change*. IPCC.

Paika, V., & Bhambri, P. (2013). Edge Detection Technique based on Fuzzy Logic. *International Journal of Mathematics and Computer Research*, *1*(3), 83–87.

Prerna, B. (2012). P., & Gupta, O.P. Multiple Sequence Alignment of Different Species. *Indian Journal of Applied Research*, *1*(7), 78–82. doi:10.15373/2249555X/APR2012/24

Reser, J. P., & Swim, J. K. (2011). Adapting to and Coping with the Threat and Impacts of Climate Change. *The American Psychologist*, *66*(4), 277–289. doi:10.1037/a0023412 PMID:21553953

Singh, S., Kakkar, P., & Bhambri, P. (2013). A Study of the Impact of Random Waypoint and Vector Mobility Models on Various Routing Protocols in MANET. *International Journal of Advances in Computing and Information Technology*, *2*(3), 41–51.

Solomon, S., Plattner, G. K., Knutti, R., & Friedlingstein, P. (2009). Irreversible Climate Change Due to Carbon Dioxide Emissions. *Proceedings of the National Academy of Sciences of the United States of America*, *106*(6), 1704–1709. doi:10.1073/pnas.0812721106 PMID:19179281

U.S. Global Change Research Program. (2018). *Impacts, Risks, and Adaptation in the United States: Fourth National Climate Assessment* (Vol. II). Retrieved from https://nca2018.globalchange.gov/

van der Linden, S. L., Leiserowitz, A. A., Feinberg, G. D., & Maibach, E. W. (2015). How to Communicate the Scientific Consensus on Climate Change: Plain Facts, Pie Charts or Metaphors? *Climatic Change*, *133*(3), 429–443.

Weber, E. U. (2010). What Shapes Perceptions of Climate Change? *Wiley Interdisciplinary Reviews: Climate Change*, *1*(3), 332–342. doi:10.1002/wcc.41

Chapter 13
Climate Change-Induced Migration and Mental Health or Displacement

Korab Jashari
Rochester Institute of Technology, Kosovo

K. Rajesh Rao
Acharya Institute of Graduate Studies, India

ABSTRACT

For the majority of people, moving abroad is already a difficult decision that brings on worry. Unfortunately, the consequences of extreme climate change are forcing people to migrate throughout the globe, causing suffering. Consequences include soil erosion, natural disasters, sea level rise, and agricultural damage. An unstoppable chain reaction is sparked by these impacts. It starts by causing environmental issues, which grow into biological issues, which directly and indirectly impact people's bodies, minds, and social structures. Furthermore, if nothing is done to stop it, this kind of displacement isn't decreasing at all and will just continue to rise. This chapter aims to provide accurate information about migration brought on by climate change, with a focus on the psychological effects; however, it also briefly discusses the physical and socioeconomic ones.

INTRODUCTION

Moving from one place to another for a better future has its drawbacks. There are many reasons why someone might want to move from home to an uncharted land. From educational purposes to political and environmental changes, people always wanted what was best for them and their families (A H Haugen, et al. 2022). In

DOI: 10.4018/979-8-3693-2177-5.ch013

this paper, we are going to focus on environmental changes and their correlation to migration, specifically caused by climate change. Meanwhile, the purpose of the paper is to shine light on the problems that climate change-induced migration causes on the human mind, body, and social elements; and try to create a cure. The paper will consist of graphs, data, and specific problems ranging from family problems, trauma, discrimination, mental health support problems, global responsibility, anxiety, and grief. The unpredictability of climate change affects even realms outside of climate-induced migration, such as environmentally induced migrations (The International Organization for Migration definition: persons or groups of persons who, predominantly for reasons of sudden or progressive change in the environment that adversely affect their lives or living conditions, are obliged to leave their habitual homes or choose to do so, either temporarily or permanently, and who move either within their country or abroad) (S, James M., et al. 2018). Because of this, people face many difficulties like social losses and the exposure of migrants to trauma and violence (S, James M., et al. 2018). It is a known fact that climate change acts in a complex and compounding manner (L. F. Walter, et al. 2022). This compound nature damages many aspects of the environment, causing environmental-induced migration in the process. Some examples of this are rising sea levels, rising ocean and land temperatures, higher disaster risks, and acceleration of rural-to-urban transplantation. Rising sea levels happen due to the melting of polar ice from increasing temperatures, leading to the reduction of habitable land surface which damages coastlines, beaches, and especially islands such as the world's 57 small island developing states. Meanwhile, rising ocean and land temperatures happen due to excessive carbon exposure in the atmosphere and other smaller variables that in the end create desertification, droughts, and drying trends which lead to food insecurity and public health crises. With ocean temperatures increasing, it changes the living, geographical, and migration patterns of fish and other organism populations. We can also draw parallels from these cases to land-living creatures, with humans included. When oceans heat most wildlife also tends to migrate to other locations, just as humans do. Meanwhile, we also have higher disaster risks like typhoons, floods, tornados, etc. That destroys human habitats and resources (Podder, S. K. et al. 2022). Lastly, the acceleration of rural-to-urban transplantation increases depression and causes mental problems, due to migrants having to face effects such as crowding, poverty, ghettoization, gang violence, and even competitive disadvantages due to a lack of urban employment skills.

All of these aspects cause enough harm to the human mind on their own, but now let's imagine what happens to these people once they reach their destination of migration. There is a ton of research and evidence that says that migrants face ridicule and discrimination after moving to these other countries. Thus, Increasing the likelihood of developing depression and having anxiety attacks due to these everyday social inequalities (Dash, L. et al. 2022). Furthermore, climate-induced natural disasters can also cause trauma to the individual which then produces stress, psychological distress, and progress to mental disorders. There are also other effects such as the exposure to loss. That be it either in material losses such as vehicles, land, or home but also social losses such as the displacement from one's community or family. This especially can cause harm to already psychologically sensitive people since it can be a catalyst to immense loneliness which later on may turn into depression or other severe mental disorders. After the departure, many people become homesick and want to return, which in some cases is impossible due to personal and social needs. This causes tremendous loneliness and the constant stress of trying to fit in (Dahal, R. et al.2022). For these reasons, the psychological perspective views this as a multiple-phase process with ever-changing stressors and not just as an episode of the human experience (Kartha, J. R. et al. 2022).

CLIMATE CHANGE'S CHAIN EFFECT

Climate change has both a domino effect and a butterfly effect. Domino meaning: it might cause some small effect on the socio-economic scale that slowly develops into a huge problem for human health. Butterfly meaning: every small change, can completely alter the results. This can especially be seen in women and children since they are more prone to getting affected by climactic events. In cases where droughts rise and the amount of nearby water decreases, it might seem like not a lot has changed and it is manageable. But this kicks off a series of domino events that can escalate very quickly if nothing is done about it. Since water is less accessible people begin to resort to sharing it, which creates a ton of gender inequalities that spiral into chaos (Rimal, Y. et al. 2022). If the situation becomes worse than it already is, then that can change how bacterial and micro-organism wildlife behaves, thus causing an eruption of infectious diseases. These changes that disturb temperature, humidity, and precipitation not only affect the world of the small but also human biology. This outcome can especially damage women's reproductive system and can cause hormonal disbalance along with it. These hormonal changes and health issues can chain effect into many mental health problems such as depression, and anxiety. In some cases, it might even cause suicidal thoughts or insanity if the problem escalates further (Kartha, J. R. et al. 2022).

Particularly noticeable are the effects of climate change in parts of the world like Western Africa. which is experiencing greater damage from climate change than ever before because of its unique location and already hot environment. Furthermore, average temperatures have been seen to have increased by 0.3 to 1.0°C, which is expected to compound over time and quicken the compounding process. The people of Africa now face increasingly harder living conditions as a result of these trends, which typically pushes them to decide to move in advance (Charanarur, P. et al.2023)). As a result, the most recent study points to migration as an adaptation to these challenging circumstances. Research even goes so far as to say that migration is not always problematic. Since the dawn of time, people have relocated for a number of advantageous reasons, including economic position, cultural interests, education, and other factors. However, when a person decides to migrate due to worsening living conditions, migration might become troublesome (Sahu, S. K. et al.2023). Clinical practices and medical education campaigns have been used by doctors and other health professionals in the UK to combat climate change. In order to improve their physical and emotional health, they have begun to advocate for less red meat consumption and more active travel. Additionally, this modification would lower greenhouse gas emissions. There is no doubt that mankind as we know it will be in danger, as well as human mental health, if something is not done to reduce the existing emissions (Garg, M. et al.2023).

CLIMATE-CHANGE MIGRATION EXAMPLES

Bangladesh has had a very difficult time since it is one of the most vulnerable countries in the world to climate change. With temperatures increasing drastically each year, it has caused damage to food security by destroying fish habitats and causing droughts to crops (Pareek, S. et al.2022). Below are some images taken from the "Climate-induced migration and population dynamics of Bangladesh" presentation by Muhammad Abdur Rahaman that represent statistics on temperature increase and the decrease of rainfall over the past 30 years in places like Patuakhali & and Barguna which are figures 1-4. We also have Table 1 which represents Disaster-affected households in specific study areas(2015-2020).

CONCLUSION

Being among the most climate change-vulnerable nations in the world, Bangladesh has had a really tough time. Finally, we may draw the conclusion that a great deal more research is required to fully understand how climate change affects mental health, given the volume of data we have gathered from various research publications. It is evident that climatic difficulties have a great deal of unintended consequences. In particular, adverse impacts like PTSD, trauma, sadness, etc. that persist over time and aren't transient issues. Of course, there are ways to lessen

Figure 1. Annual average temperatures Patuakhali and Barguna (1990-2020)

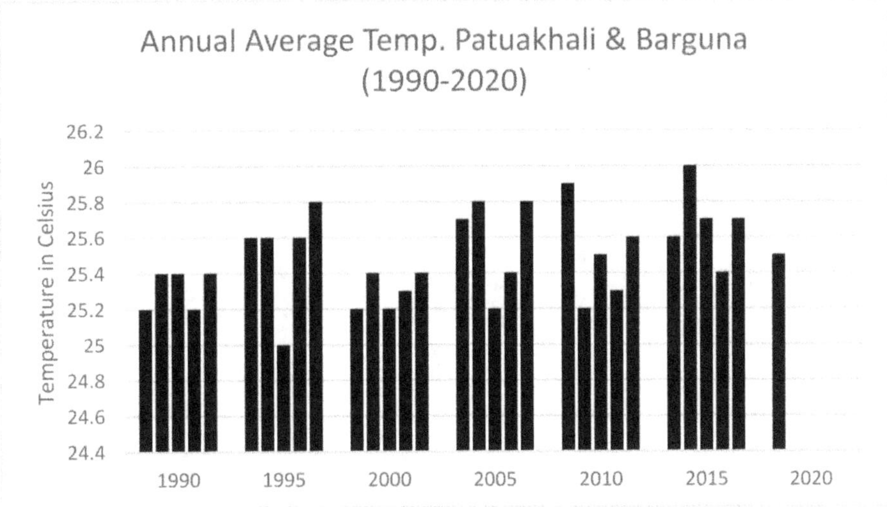

Figure 2. Annual average temperatures Satkhira (1990-2020)

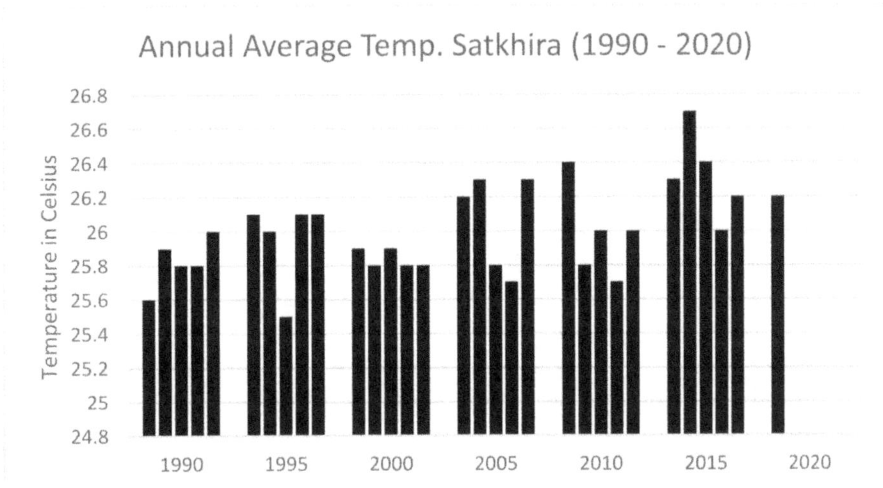

the harm, such as requiring physicians and medical facilities to educate patients and providing them with specialized resources that make it simpler for them to spread the word than it is now. Considering this demonstrates that, except than purchasing time for new technologies to develop, there is little that can be done

Figure 3. Annual average rainfall Patuakhali and Barguna (1990-2020)

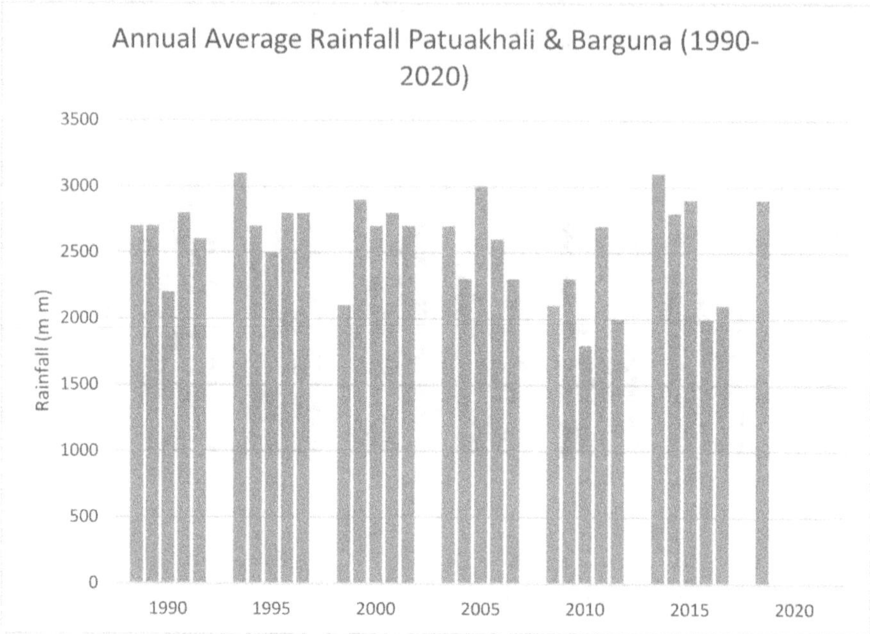

Figure 4. Annual average rainfall Patuakhali and Barguna (1990-2020)

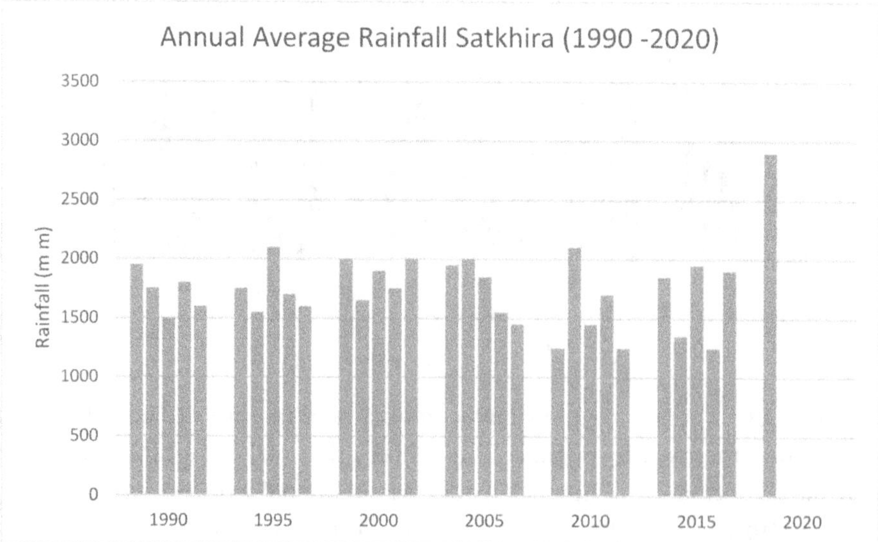

Table 1. Disaster-affected households in the study area (2015-2020)

Disaster-Affected Households in the Study Area (2015-2020)					
Name of disaster	**Patuakhali**	**Barguna**	**Satkhira**	**Rajshahi**	**Chapai Nawabagnj**
Drought	693	-	240	2,014	3,426
Flood	62,136	-	6,560	82,597	11,644
Water Logging	27,117	16,865	35,442	8,910	32
Cyclone	1,03,488	1,12,163	2,12,024	38,970	411
Tornado	-	11,366	425	5,813	11
Storm and tidal Surge	19,674	95	2,181	-	-
Thunderstorm and Lighting	1,231	1,365	3,896	16,454	22,932
River and coastal Erosion	6,250	268	11,231	11,131	10,719
Salinity	-	-	4,10,020	-	-
Hailstorm	868	740	6,239	-	-

to combat climate change. Because evidence indicates that the effects of climate change are happening very quickly right now and because we are currently facing one of the worst foes that humanity has ever faced. In light of this, although there is now no cure for climate change, we can yet hold out hope that perhaps global reforms will occur quickly enough to address this enormous issue.

REFERENCES

Ahmad, Z., Maskat, R., & Mohamed, A. (2023). Harnessing Natural Language Processing for Mental Health Detection in Malay Text: A Review. *2023 4th International Conference on Artificial Intelligence and Data Sciences (AiDAS),* 29-35. 10.1109/AiDAS60501.2023.10284653

Ajith, M., & Calhoun, V. D. (2023). Functional Network Connectivity Based Mental Health Category Prediction from Rest-fMRI Data. *2023 IEEE 20th International Symposium on Biomedical Imaging (ISBI),* 1-5. 10.1109/ISBI53787.2023.10230721

Akbar, D. A., Bewasana, D. A. D., Mahfud, A. Z., & Yulita, T. (2023). InnerCheck: A New Breakthrough for Early Detection of Mental Health. *2023 IEEE International Conference on Cryptography, Informatics, and Cybersecurity (ICoCICs),* 281-286. 10.1109/ICoCICs58778.2023.10276558

Almaleh, A. (2023). Machine Learning-Based Forecasting of Mental Health Issues Among Employees in the Workplace. *2023 IEEE International Conference on Industry 4.0, Artificial Intelligence, and Communications Technology (IAICT),* 118-124. 10.1109/IAICT59002.2023.10205620

Anand, N., Pant, L. M., Alam, T., Pundir, S., Thomas, L., & Rakshith, U. R. (2023). Artificial Intelligence's Contribution To Mental Health Education. *2023 International Conference on Sustainable Computing and Data Communication Systems (ICSCDS),* 462-467. 10.1109/ICSCDS56580.2023.10104905

Askland. (2022). Beyond Migration: A Critical Review of Climate Change Induced Displacement. *Environmental Sociology, 8*(3), 267–278. . doi:10.1080/23251042.2022.2042888

Bai, L., & Han, X. (2023). Design of A Mental Health Assessment System Based on Fuzzy Evaluation. *2023 International Conference on Data Science and Network Security (ICDSNS),* 1-6. 10.1109/ICDSNS58469.2023.10245787

Bajaj, V., Bathija, R., Megnani, C., Sawara, J., & Ansari, N. (2023). Non-Invasive Mental Health Prediction using Machine Learning: An Exploration of Algorithms and Accuracy. *2023 7th International Conference on Intelligent Computing and Control Systems (ICICCS),* 313-321. 10.1109/ICICCS56967.2023.10142504

Brocki, L., Dyer, G. C., Gładka, A., & Chung, N. C. (2023). Deep Learning Mental Health Dialogue System. *2023 IEEE International Conference on Big Data and Smart Computing (BigComp),* 395-398. 10.1109/BigComp57234.2023.00097

Brown. (2022). Mental Health Impacts of Climate Change: Considerations for Nurse Practitioners. *The Journal for Nurse Practitioners, 18*(4), 359–363. . doi:10.1016/j.nurpra.2021.07.013

Ch, Dixit, Joshi, Pant, Thomas, & Beri. (2023). Predicting Employee Mental Health Using Artificial Neural Network. *2023 3rd International Conference on Advance Computing and Innovative Technologies in Engineering (ICACITE),* 1380-1383. 10.1109/ICACITE57410.2023.10183130

Charanarur, P., Jain, H., Rao, G. S., Samanta, D., Sengar, S. S., & Hewage, C. T. (2023). Machine-Learning-Based Spam Mail Detector. *SN Computer Science, 4*(6), 858. doi:10.1007/s42979-023-02330-x

Chen, T., Song, L., Zhou, H., Li, Y., Wang, H., & Kong, C. (2023). A Study of Mental Health Self-Monitoring Based on the Combination of BERT and Low-Code Platform. *2023 6th World Conference on Computing and Communication Technologies (WCCCT),* 171-175. 10.1109/WCCCT56755.2023.10052392

Dalvi, A., Kapur, P., Ashtekar, R., Gorugantu, A., & Siddavatam, I. (2023). Transforming College Counseling: An In-House Solution for Student Mental Wellness. *2023 2nd International Conference for Innovation in Technology (INOCON),* 1-5. 10.1109/INOCON57975.2023.10101364

Dash, L., Pattanayak, B. K., Singh, D., Samanta, D., & Rimal, Y. (2022). Network Lifetime Enhancement by Elimination of Spatially and Temporally Correlated RFID Surveillance Data in WSNs. *Wireless Communications & Mobile Computing.*

Fareesha. (2023). Real-Time Artificial Mood-tracking and Health-monitoring System (RAMAHS) for people with mental illness and their Caregivers. *2023 International Conference on Recent Trends in Electronics and Communication (ICRTEC),* 1-6. 10.1109/ICRTEC56977.2023.10111897

Filho. (2022). Where to Go? Migration and Climate Change Response in West Africa. *Geoforum, 137,* 83–87. . doi:10.1016/j.geoforum.2022.10.011

Giovanis, E., & Abdeldjalil, A A. (2023). The Impact of Climate Change-Induced Migration on Well-being In Joint Policy Study. *Climate-Induced Internal Displacement in MENA Countries.*

Grasser, L. R. (2022). Addressing Mental Health Concerns in Refugees and Displaced Populations: Is Enough Being Done? *Risk Management and Healthcare Policy, 15,* 909–922. . doi:10.2147/RMHP.S270233

Gupta, V., Joshi, V., Jain, A., & Garg, I. (2023). Chatbot for Mental health support using NLP. *2023 4th International Conference for Emerging Technology (INCET),* 1-6. 10.1109/INCET57972.2023.10170573

Jacqueline, M. (2017). *The centrality of social ties to climate migration and mental health. BMC Public Health,* 17. doi:10.1186/s12889-017-4508-0

Jage, S., Chaudhari, S., Jatte, M., Mhatre, A., & Mane, V. (2023). Predicting Mental Health Illness using Machine Learning. *2023 3rd Asian Conference on Innovation in Technology (ASIANCON),* 1-5. 10.1109/ASIANCON58793.2023.10270445

Jones, A. (2022). The Health Impacts of Climate Change: Why Climate Action Is Essential to Protect Health. *Orthopaedics and Trauma, 36*(5), 248–255. doi:10.1016/j.mporth.2022.07.001

Jusoh, S., Kadir, R. A., & Al Fawareh, H. (2023). mentBot: A Repository for Mental Health Disorders. *2023 International Conference on Consumer Electronics - Taiwan (ICCE-Taiwan),* 153-154. 10.1109/ICCE-Taiwan58799.2023.10227068

Karthi, M., Alsager, M., Metha, R., Fatima, N. N., & Al-Gindy, A. (2023). Digital Solution: Breaking the Barriers to Address Stigma of Mental Health. *IEEE EUROCON 2023 - 20th International Conference on Smart Technologies*, 24-29. 10.1109/EUROCON56442.2023.10198879

Kumar, K., Chaudhary, S., Anandaram, H., Kumar, R., Gupta, A., & Joshi, K. (2023). Industry 4.0 and Health Care System with special reference to Mental Health. *2023 1st International Conference on Intelligent Computing and Research Trends (ICRT),* 1-5. 10.1109/ICRT57042.2023.10146640

Li, Y. (2023). Application of Machine Learning to Predict Mental Health Disorders and Interpret Feature Importance. *2023 3rd International Symposium on Computer Technology and Information Science (ISCTIS),* 257-261. 10.1109/ISCTIS58954.2023.10213032

Liao. (2023). Association between Internal Migration Experience and Depressive Symptoms: Analysis of PSID Data. *BMC Public Health, 23*(1). . doi:10.1186/s12889-023-16073-0

Lohia, A., Ranjith, A., Sachdeva, A., & Bhat, A. (2023). An Ensemble Technique to Analyse Mental Health of Workforce in the Corporate Sector. *2023 7th International Conference on Intelligent Computing and Control Systems (ICICCS),* 1245-1251. 10.1109/ICICCS56967.2023.10142470

Lokotola, C. L. (2023). Towards a climate-resilient primary health care service. *South African Family Practice, 65*(1). . doi:10.4102/safp.v65i1.5749

Marandi, A. K., & Shah, H. (2023). Treatment of Mental Health Disorders Advanced By Artificial Intelligence. *2023 3rd International Conference on Advance Computing and Innovative Technologies in Engineering (ICACITE),* 2569-2573. 10.1109/ICACITE57410.2023.10182496

Marazziti. (2021). Climate Change, Environment Pollution, COVID-19 Pandemic and Mental Health. *Science of The Total Environment, 773*, 145182. doi:10.1016/j.scitotenv.2021.145182

McIver. (2016). Health Impacts of Climate Change in Pacific Island Countries: A Regional Assessment of Vulnerabilities and Adaptation Priorities. *Environmental Health Perspectives, 124*(11), 1707–1714. doi:10.1289/ehp.1509756

Nadeem, M., Rashid, J., Moon, H., & Dosset, A. (2023). Machine Learning for Mental Health: A Systematic Study of Seven Approaches for Detecting Mental Disorders. *2023 International Technical Conference on Circuits/Systems, Computers, and Communications (ITC-CSCC),* 1-6. 10.1109/ITC-CSCC58803.2023.10212609

Nagarajaiah, K., Krishnappa, M. H., & Asha, K. R. (2023). Machine Learning based Detection of Post Traumatic Stress Disorder of Mental Health. *2023 4th International Conference on Electronics and Sustainable Communication Systems (ICESC),* 1456-1460. 10.1109/ICESC57686.2023.10193036

Nicholas. (2020). Mental Health Impacts of Climate Change: Perspectives for the ED Clinician. *Journal of Emergency Nursing, 46*(5), 590–599. doi:10.1016/j.jen.2020.05.014

Nickerson. (2019). The Association between Visa Insecurity and Mental Health, Disability and Social Engagement in Refugees Living in Australia. *European Journal of Psychotraumatology, 10*(1). doi:10.1080/20008198.2019.1688129

Nouman, M., Sara, H., Khoo, S. Y., Mahmud, M. P., & Kouzani, A. Z. (2023). Mental Health Prediction through Text Chat Conversations. *2023 International Joint Conference on Neural Networks (IJCNN),* 1-6. 10.1109/IJCNN54540.2023.10191849

Osintseva, A. A., & Gvozdetskaya, S. S. (2023). ICT and Mental Health of Professionals Working from Home. *2023 IEEE 24th International Conference of Young Professionals in Electron Devices and Materials (EDM),* 1980-1984. 10.1109/EDM58354.2023.10225054

Podder, S. K., Thomas, B., & Samanta, D. (2022). Impact of Innovative Technology on Quality Education: The Resourceful Intelligence for Smart Society Development. *Machine Intelligence and Data Science Applications: Proceedings of MIDAS 2021, 132*, 293.

Poursadeqiyan. (2020). Migration Health Crisis Associated with Climate Change: A Systematic Review. *Journal of Education and Health Promotion, 9*(1), 97. doi:10.4103/jehp.jehp_4_20

RahamanM. A. (2023). *Climate Induced Migration and Population Dynamics of Bangladesh*. doi:10.13140/RG.2.2.13575.98729

Ramadan & Ataallah. (2021). Are Climate Change and Mental Health Correlated? *General Psychiatry, 34*(6). . doi:10.1136/gpsych-2021-100648

Rani, K., Vishnoi, H., & Mishra, M. (2023). A Mental Health Chatbot Delivering Cognitive Behavior Therapy and Remote Health Monitoring Using NLP And AI. *2023 International Conference on Disruptive Technologies (ICDT),* 313-317. 10.1109/ICDT57929.2023.10150665

Rimal, Y., Rathore, V. S., Pageni, S., Samanta, D., Karuppiah, M., & Singh, D. (2022). Neural Network based Student Grade Prediction Model. *2022 International Conference on Machine Learning, Computer Systems and Security (MLCSS)*, 22–27. 10.1109/MLCSS57186.2022.00013

Sahu, S. K., Chandra, J., Muloor, K., & Samanta, D. (2023). A Thorough Investigation of Various Goals and Responses for Mobile Software-Defined Networks. In *Designing and Developing Innovative Mobile Applications* (pp. 307–328). IGI Global. 10.4018/978-1-6684-8582-8.ch017

Salik, K. M. (2023). *A Threat or an Opportunity? Internal Migration in the Context of Climate Extremes in Pakistan.* doi:10.1007/978-3-031-34194-6_11

Samuel, M., & Shirley, C. P. (2023). Mindset, An Android-Based Mental Wellbeing Support Mobile Application. *2023 3rd International Conference on Pervasive Computing and Social Networking (ICPCSN)*, 989-996. 10.1109/ICPCSN58827.2023.00169

Schwerdtle. (2018). The Health Impacts of Climate-Related Migration. *BMC Medicine, 16*(1). . doi:10.1186/s12916-017-0981-7

Sheffield & Landrigan. (2011). Global Climate Change and Children's Health: Threats and Strategies for Prevention. *Environmental Health Perspectives, 119*(3), 291–298. doi:10.1289/ehp.1002233

Shultz. (2018). Public Health and Mental Health Implications of Environmentally Induced Forced Migration. *Disaster Medicine and Public Health Preparedness, 13*(2), 116–122. doi:10.1017/dmp.2018.27

Sri. (2023). Global Mental Health and Climate Change: A Geo-Psychiatry Perspective. *Asian Journal of Psychiatry, 84*, 103562. . doi:10.1016/j.ajp.2023.103562

Uddin. (2021). A Global Child Health Perspective on Climate Change, Migration and Human Rights. *Current Problems in Pediatric and Adolescent Health Care, 51*(6). . doi:10.1016/j.cppeds.2021.101029

United States Congress, Climate Change Science Program Office. (2008). *Analyses of the Effects of Global Change on Human Health and Welfare and Human Systems: Final Report, Synthesis and Assessment Product 4.6: Report by the U.S. Climate Change Science Program and the Subcommittee on Global Change Research.* U.S. Climate Change Science Program.

Vafeiadou. (2023). The Influence of Climate Change on Mental Health in Populations of the Western Pacific Region: An Umbrella Scoping Review. *Heliyon, 9*(11). . doi:10.1016/j.heliyon.2023.e21457

Virani, J., Daredi, N., Bhanushali, A., Shukla, M., & Shah, P. (2023). Mental Healthcare Analysis using Power BI & Machine Learning. *2023 4th International Conference on Signal Processing and Communication (ICSPC),* 73-76. 10.1109/ICSPC57692.2023.10125680

Wright. (2023). Climate Change and the Adverse Impact on the Health and Well-Being of Women and Girls from the Women's Health Expert Panel of the American Academy of Nursing. *Nursing Outlook, 71*(2). . doi:10.1016/j.outlook.2023.101919

Zhu, L., & Zhang, Y. (2023). Fuzzy Judgment-based Mental Health Assessment System for College Students. *2023 IEEE International Conference on Integrated Circuits and Communication Systems (ICICACS),* 1-5. 10.1109/ICICACS57338.2023.10099884

Zim, M. K. I., & Kaur, U. (2023). Mental health detection and anticipation analysis using Deep Learning. *2023 International Conference on Computer Communication and Informatics (ICCCI),* 1-6. 10.1109/ICCCI56745.2023.10128581

Chapter 14
Climate Change:
A Threat to Economic Stability and Equality

Rona Hoxha
Rochester Institute of Technology, Kosovo

A. R. Deepti
Acharya Institute of Graduate Studies, India

ABSTRACT

This chapter explains how the financial stability of the economy is threatened by the effects of climate change. Since research and statistics on the subject are still deemed to be shallow and inadequate, the purpose of this chapter is to address the lack of understanding regarding the extent of the effects that climate change poses to the financial industry. A wealth of information about the many economic sectors and systems impacted by climate change can be gained by investigating several economic models, such as GEM-E3-ISL and GINFORS.

METHODOLOGY AND MODELS USED FOR THE SITUATIONAL ANALYSIS

For assessing the possible economic repercussions, Computable General Equilibrium (CGE) models and Integrated Assessment models (IAM) are used. The main difference of CGE and IAM models, is that CGE models aim to analyze the relationship between different economic stakeholders and its domains. In order to address climate change outcomes, it measures the loss on the markets, i.e., altering the production of crops, health expenses, labor capacity and energy use in economic enterprises (Medani P.

DOI: 10.4018/979-8-3693-2177-5.ch014

Bhandari, 2021). While IAMs, core element lays on representing the collaboration between the economic and biophysical dimension. Each one of these models contain their own setbacks, making it imperative to implement both into one new enhanced model, in order to yield the most productive and successful result yet. Such model exists, called the ENV-Linkages. Besides its elevated elements incorporated into the model, ENV-Linkages also implement environmental pressure into the description, specifically greenhouse gas emissions (GHG). It serves a connective bridge between the economic and impact model, feeding the input from the previous models, back into the CGE model.

GEM-E3-ISL Model

A CGE model tailored to a particular nation, the GEM-E3-ISL model functions across various economic sectors. This model includes economic actors operating at the subnational level, such as households, businesses, governments, and enterprises (Medani P. Bhandari; Shvindina Hanna, 2021). It is an active, self-repeating model. Since islands' economies are small and unaffected by global financial issues, this approach was found to be more appropriate.

A spokesperson is presumed from among the several production areas that GEM-E3-ISL takes into account. While business owners raise their profits, consumers work to improve their well-being as well. Another supposition is that marketplaces eventually become competitive, leading to price takers among manufacturers and consumers (Medani P. Bhandari; Shvindina Hanna, 2021). The initial stock of productive capital, which consists of the required equipment and instruments, may vary as new enterprises are undertaken (Yuliia Matvieieva et al, 2021).This is the previously described dynamic and active aspect. As a result, more money will be allocated to that product when various businesses observe variations in demand. The notion that involuntary unemployment may prevent some resources—people who wish to work, primarily—from being utilized in the labor market is taken into consideration. This is in line with the idea of "efficiency wages," according to which the availability of labor and pay have an impact on the employment prospects of job seekers.

According to the models closer rule, an induvial's savings will always equal the amount invested over time. This is accomplished by altering the local interest rate, demonstrating the disparities in the financial expenses faced by enterprises in various locations (Medani P. Bhandari et al, 2021). A shock that is caused by demand, such as a shift in tourism earnings, affects not only the activities in the immediate sector but also the activities in secondary sectors (P. Pepe, 2022). Because of market competitiveness and the financial ramifications of climate change, resources of the economy concentrate on producing more capital through the activities that produce the most advantageous outcomes when changes in production costs occur (Podder, S. K. et al. 2022). The economy's primary production components, labor and capital, are fixed at the regional level because this model is essentially country-specific, which means that islands are limited by their own capitals, assets, and finances. According to W. Gu et al. (2017), this assumption limits the capacity for labor movement and outside investment, which hinders adaptation and growth and may have negative economic effects.

A METHODOLOGY FOR ESTIMATING CLIMATE DAMAGES

Through certain methodologies, we can secure the overall number of economic costs from climate change on gross domestic product of a country. Comprehensive GDP expenses portray the degree of the consequences' severity from climate change. Which will potentially hinder future economic development. This concludes that, damages substantially affect key components of the economy and the economy's development rate 9 Dash, L. et al.2022).

Two main types of climate change fallouts are evident. They are classified as: The supply and the demand side of the economic mechanism.

First Category, Supply Side

For the agriculture sector, outcomes have been tracked with exogenous sector, showing that changes in crop growth ratio are interlinked with how other factors such as qualitative soil hinders or helps these practices. For the fisheries sector, outcomes have been tracked through reduction of the stock, which resembles the aftermath of climate change for the biodiversity and its sector. Coastal area impacts of sea level increase are mimicked through reductions in the accessible land and destruction of physical capital in specific areas. As for the ramifications of extreme weather events, only the damages from increased hurricane activity are used as an instrument, because of information insufficiency, it is assumed that this event influences all segments of the economy. Since other extreme weather activities contain substantial doubts involved with their effects, the example of tropical cyclones is applied. Figure 1 represents Shocks and Vulnerabilities Trigger Financial Stability (Dahal, R. et al.2022).

The 3 Main Conclusions

Vulnerabilities and shocks are used as instruments in forecasting possible impacts to the economic sector.

Figure 1. Shocks and vulnerabilities trigger financial stability

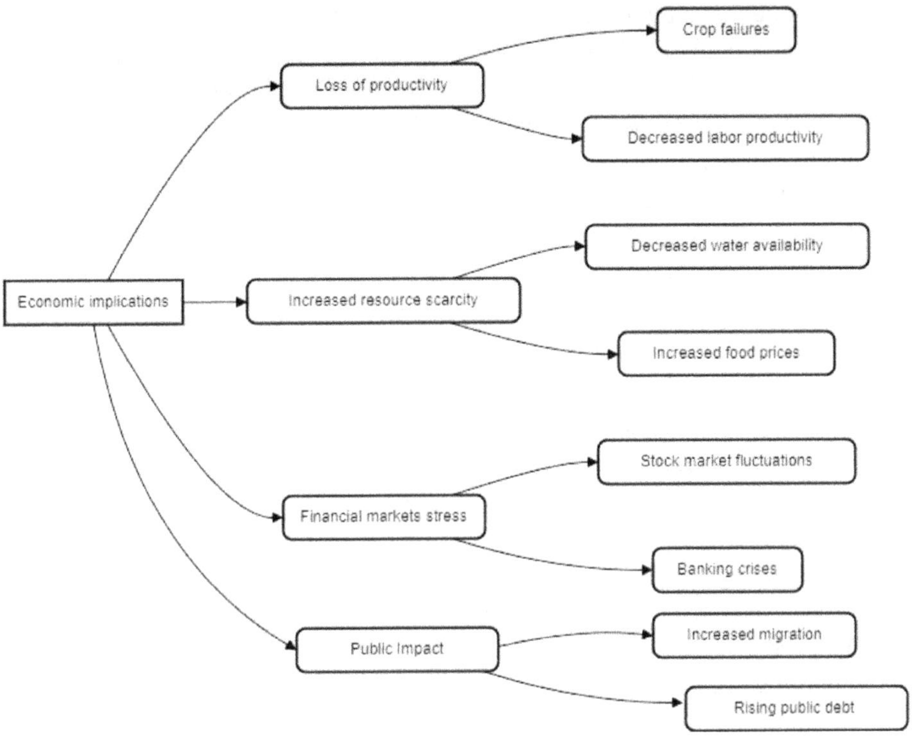

- First conclusion being, the Federal Reserve's financial stability monitoring framework is versatile enough, enabling the inclusion of vital components of climate-related implications.
- Second, although it is thought and inferred that climate change increases risks for financial stability, more data and analysis are vital for implementing financial stability monitoring (H. D. Mohammadian, V. Wittberg, M. Castro and G. Bolandian, 20xx).
- Thirdly, global transparency regarding the scope of climate change in each and every region can help deductively and precisely determine the seriousness and diffusion of climate change.

Despite the consensus among scientists regarding the patterns and triggers of climate change, the exact eruption and magnitude of climate aftermath remain unpredictable. Climate risks are defined as the various potential consequences and impacts associated with the changes of climate conditions.

Climate change risk functions demonstrate the possible effects of climate change to the economy, degenerating public health and welfare, work performance, and crop productions. While also increasing death rates and climate-related harm to assets. Such consequences conclude in direct financial risks, which subsequently may change the cost and accessibility of credit, altering the time sequence and reliability level of cash flows. Thus, resulting in risks to economic activity, which magnitudes financial risks and vice-versa (Kartha, J. R. et al.2022). Climate related or financial risks as an individual instrument cannot trigger financial stability, although in the case when these implications cause financial-system vulnerabilities to fluctuate through revenue losses of financial traders, "hiccups" in the operation of the financial system, or abrupt repricing of significant asset classes.

This overflow of outcomes from climate change fits into one out of two ways from the monitoring framework within the Federal Reserve. First, these climate threats can emerge as shocks to the financial system. Unpredictable severe weather events could potentially trigger the stability of financial assets or their value(Sahu, S. K. et al.2023). This poses a risk to investors and their expectations. Second being the increase of financial system vulnerabilities which, could boost shock. Opacity of exposures, mispricing and correlated aggregate exposure are the examples that escalate the chance of asset mispricing and other unfortunate financial situations (Rimal, Y. et al.2022).

- **Opacity of exposures:** The geographic distribution of many supply chains is not disclosed neither tracked, there's scarce information about the detailed results from climate change in financial and non-financial enterprises as well. Lack of adoption and implementation on models and frameworks concerning climate change risks, creates easy access for vulnerabilities and shocks to financial stability. Which leads to amplified repricing.
- **Mispricing:** Research demonstrates that even though investors in some cases are aware and informed about possible climate risks, they underestimate its destructive outcomes, resulting in physical risks

Correlated aggregate exposure. Another significant path would be the possibility of non-linearities so, situations where the impact may not increase steadily with the magnitude of changes; there could be sudden and unpredictable shifts or, points where the climate system may undergo abrupt and substantial changes may happen.

MODEL CALIBRATION

The No-Damage Socioeconomic Baseline

Examinations expected expenses from climate change relies on a baseline projection known as a "no damage" assumption. This scenario assumes that new climate policies are not integrated, although, the continuation of existing policies remains implemented and practiced. It is vital to consider that this baseline projection doesn't predict the future; instead, it produces a possible storyline of how key trends might impact economic development without unexpected shocks to the financial stability and overall economy.

UN 2013 and EUROSTAT 2013 provide accurate demographic trends through taking into consideration the population's age, gender and education. Demographic patterns act as a crucial factor impacting the development of economy. From there, this data is gathered. It is expected that the population reaches nearly 10 billion people by 2060, in a global level. Despite this overall percentage of increase, the growth rate is estimated to be evidently lower by 2050 compared to current levels in many regions of the globe(H. D. Mohammadian, 2022). The correlation between population and employment varies across parts of the world, reflecting possible results about employment and participation quo across different age groups.

Possible projections for the regional GDP reveal that economic growth tolls are progressively decreasing across multiple regions, while still remaining possible. The most significant progress is expected to be particularly in the Afro-Asian world, which are considered to be areas with extensive economic potential. These projections and assumptions align with the OECD Economic Outlook (J. Vishnu et al., 2014). GDP growth is a result of 3 products intertwined. First, increased supply of production factors, second would be, geographical changes of resources, and lastly, advancing effectiveness and efficiency of resource usage. In the long run, there is a shift in the direction of a more stable growth path where workforce productivity, being the key components that drives economic progress, is matched by grander capital supply(Garg, M. et al.2023).

To better conceptualize how the future economy is going to look like, the focus shouldn't just be in the bigger picture; various aspects need to be considered for s better estimate. One example would be how much productivity is being put into each economic sector, and how much capital individuals generate in each area. As the consumers income amplifies, intentions and expectations may change(Charanarur, P. et al.2023). Thus, altering international relations and their market trading, maintaining stability. MENA countries heavily rely on oil trading for primary generation of income in their economies, are expected to open their economies to new diverse option for market, with the intention of eventually lowering usage of energy. In developing countries, the significance of agriculture and its practice is believed to reduce furthermore, because the target centered is centered more towards industrial and service-related activities, especially with the high progression rate(Pareek, S. et al.2022).

Looking at energy prediction until 2035, based on the International; Energy Agency (IEA) estimates will happen through current implemented policies. They resulted in their own scientific projections on the potential world economy status. It is calculated that countries like China, India and Indonesia which are developing quickly, might start using coal, a mineral that will aid their economy to a greater extent. While in the United States, the utilization of natural gas is set to rise. Figure 2 represents the No-Damage Socioeconomic Baseline.

Figure 2. The no-damage socioeconomic baseline

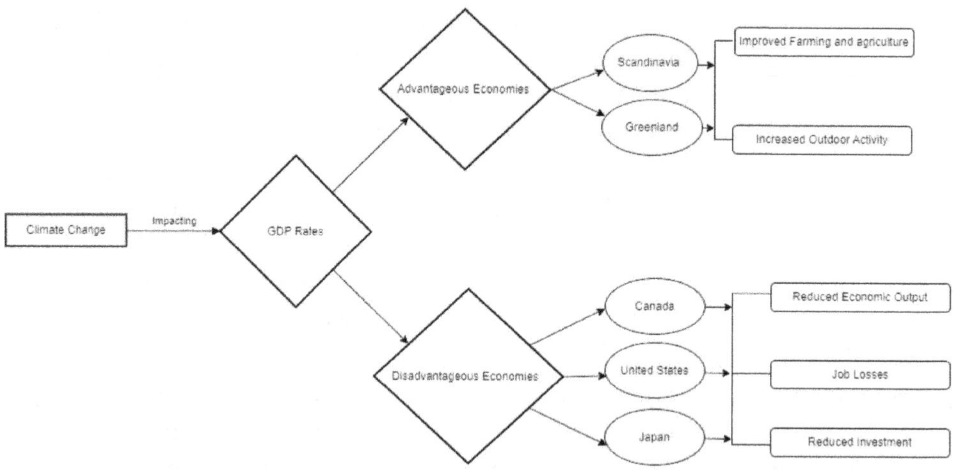

Greenhouse Gas Emissions and Climate Factors

Fundamental economic events contribute to a continuous rise in local and global emissions. Predictions indicate that global human-made greenhouse gas (GHG) emissions are expected to climb from approximately 45 billion metric tons (Gt) of CO2 equivalent (CO2e) in 2010 to about 95 GtCO2e in 2060. Carbon dioxide (CO2) is projected to continue to be the predominant greenhouse gas. The rise in emissions aligns with significant demographic forecasts, including bigger populations, heightened economic activities, and increased utilization of fossil fuels. Regardless of the decrease in population and GDP rates, the shift in economic prominence toward rising economies, coupled with the continues extensive use of fossil fuel since new climate policies are yet to be developed, leads to substantial surge in GHG emissions (Ippolito, M., & Cicatiello, L. 2019). Notably, the magnified use of coal, as detailed earlier, speeds up emission increases. Nevertheless, there is a degree of relative dissociating: emission grow at a slower pace than production.

RESULTS ON THE ECONOMIC RAMIFICATIONS

Macroeconomic and Structural Impacts

The regional repercussions of global warming on the economic structure of all economies worldwide, but the discrepancy in regional level remains substantial. It is foreseen that global emission rates will suffer a great increase, and the probability of destructive catastrophes will intensify in approximately 40 years.

The diverse climate repercussions manifest over varying timeframes, leading to non-identical GDP fluctuations across regions. The most susceptible areas exhibit a non-linear trajectory of damages as time (and climate change) progress. This nonlinearity of regional and global damages over time reflects a convergence of several key catalysts.

The consequences come in a variety of ways, reflecting in different spans of time, resulting in GDP variation across regions. The most sensitive areas experience increases in damages as time and climate change progress (Nau, M. D. 2016). Since local temperatures continue to surge, the consequences intensify. On the contrary, temperatures patterns in a global scale, are predicted to remain undeviating, the ecological implications inflate. It is especially evident in crop losses in agriculture. This discrepancy is accurate because the economy fails to adapt to changes in factor supply and production. When insignificant shocks occur, relocation costs are minor, while for greater shocks, relocation costs magnify.

The economic sphere of nations globally, is severely affected from the ecological after-effects of climate change. They directly and indirectly influence other industries as well. An example would be the alter of production in manufacturing industries due to price changes, capital damage etc. The financial costs of global warming are not evenly distributed across regions (Galbraith, 2012). Developing countries are generally more vulnerable to climate change than developed countries. This is because developing countries tend to have a greater reliance on climate-sensitive sectors, such as agriculture, and weaker infrastructure. Additionally, developing countries often have less capacity to adjust to the implications caused by climate change. It is anticipated that the future will reveal an increased level of severity for the economic ramifications (Pauly, R, 2012). This is because, they are expected to intensify over time, substantially making the global economy increasingly interconnected and vulnerable, which means that the outcome of climate change in one region can quickly diffuse onto other neighboring regions.

Although the results of climate change present a significant task for policymakers, mitigating climate change through: lowering carbon emissions, investing in renewable energy and energy efficiency, advancing infrastructure resilience, arranging climate-related funds for developing countries, are crucial steps that should be taken in order to minimize these effects and the overall aftermath, while also helping us familiarize with the existing patterns and trends (Ali, A. 2018).

There are significant and projected to become more pronounced effects from the complex relationship between climate change and the economy. These effects include a wide range of economic aspects, each posing particular difficulties:

- Reduced Productivity: The weather extremes brought on by climate change have an impact on agriculture and industry. Events such as heatwaves, storms, and changes in precipitation patterns can cause productivity losses in a number of industries.
- Increased Resource Scarcity: Droughts and variations in precipitation are two examples of environmental changes that can cause resource scarcities. Particularly, sectors reliant on water supplies may be impacted by shortages of water, further taxing the economy.
- Financial Market Stress: Uncertainties associated to climate change add another degree of risk to the financial system. Assessing and reducing the risks related to the effects of climate change is becoming more difficult for insurers, investors, and other financial stakeholders.
- Increased Migration: Communities may be forced to migrate as a result of climate-related catastrophes like rising sea levels and harsh weather. In addition to putting a burden on social structures, this population displacement creates financial difficulties for those who must relocate and support themselves.
- Risking Public Debt: As governments respond to climate-related disasters and put adaptation plans into place, they may have to shoulder heavier financial responsibilities. This may result in high amounts of public debt, which could have an effect on the fiscal stability of a country.
- GDP Losses: The combined impact of these issues brought on by climate change may cause large GDP losses. An economic downturn is caused by lower productivity, higher resource prices, and disruptions to important industries.

CONCLUSION

Climate changes poses immense risks to the economy while also creating financial instability. Potential repercussions are already slightly evident, and it is foreseen that the implications will worsen progressively with climate change. Fluctuations in economy consist of: Loss of productivity, increased source scarcity, financial market stress, increased migration, and risking public debt, GDP losses etc.

REFERENCES

Ali, A. (2018). issue of income inequality under the perceptive of macroeconomic instability: An empirical analysis of pakistan. *Pakistan Economic and Social Review, 56*(1), 121–155.

Charanarur, P., Jain, H., Rao, G. S., Samanta, D., Sengar, S. S., & Hewage, C. T. (2023). Machine-Learning-Based Spam Mail Detector. *SN Computer Science, 4*(6), 858. doi:10.1007/s42979-023-02330-x

Dahal, R., Samanta, D., Karuppiah, M., & Biswas, J. (2022). Comparative Study of Graph Theory for Network System. International Conference on Innovative Computing and Communications. *Proceedings of ICICC, 3*, 73–84.

Dash, L., Pattanayak, B. K., Singh, D., Samanta, D., & Rimal, Y. (2022). Network Lifetime Enhancement by Elimination of Spatially and Temporally Correlated RFID Surveillance Data in WSNs. *Wireless Communications and Mobile Computing, 2022*, 1–12. doi:10.1155/2022/2437578

Galbraith, J. K., & ProQuest (Firm). (2012). *Inequality and instability: A study of the world economy just before the great crisis* (1st ed.). Oxford University Press. doi:10.1093/acprof:osobl/9780199855650.001.0001

Garg, M., Kumar, M., & Samanta, D. (2023). *Towards Pattern Recognition with Network Science and Natural Language Processing for Information Retrieval.* Academic Press.

Gu, W., Lou, G., Tan, W., & Yuan, X. (2017, November). A Nonlinear State Estimator-Based Decentralized Secondary Voltage Control Scheme for Autonomous Microgrids. *IEEE Transactions on Power Systems, 32*(6), 4794–4804. doi:10.1109/TPWRS.2017.2676181

Ippolito, M., & Cicatiello, L. (2019). Political instability, economic inequality and social conflict: The case in italy. *Panoeconomicus, 66*(3), 365–383. doi:10.2298/PAN1903365I

Kartha, J. R., Dahal, R., Samanta, D., Poonia, R. C., Bhattacharya, A., & Dutta, S. (2022). A Shortest Path Problem for Drug Delivery Using Domination and Eccentricity. In Human-Centric Smart Computing [Springer Nature Singapore Singapore.]. *Proceedings of ICHCSC, 2022*, 207–215.

Liubov Zharova. (2021). *4 Sustainable Development Goals as a Roadmap for Territorial Development," in Economic Inequality – Trends, Traps and Trade-offs.* River Publishers.

Medani, P. (2021a). *1 Trends, Traps and Trade-offs of Economic Inequality—A Growing Pattern and Concern, in Economic Inequality – Trends, Traps and Trade-offs*. River Publishers.

Medani, P. (2021b). *12 Conclusion—The Unpredictable Direction of Economic Inequality, in Economic Inequality – Trends, Traps and Trade-offs*. River Publishers.

Medani, P. (2021c). *Authors Biographies, in Economic Inequality – Trends, Traps and Trade-offs*. River Publishers.

Medani, P. (2021d). *Economic Inequality – Trends, Traps and Trade-offs, in Economic Inequality – Trends, Traps and Trade-offs*. River Publishers.

Medani, P. (2021e). *Word Phrase Definitions and Abbreviation for Economic Inequality – Trends, Traps and Trade-offs, in Economic Inequality – Trends, Traps and Trade-offs*. River Publishers.

Mohammadian, H. D., Wittberg, V., Castro, M., & Bolandian, G. (2020). *The 5th Wave and i-Sustainability Plus Theories as Solutions for SocioEdu Consequences of Covid-19. In 2020 IEEE Learning With MOOCS*. LWMOOCS., doi:10.1109/LWMOOCS50143.2020.9234360

Nau, M. D. (2016). *Whose financial crisis? how the great recession reshaped economic instability and inequality in the U.S.* Academic Press.

Pareek, S., Bhatnagar, V., Poonia, R. C., Sharma, S., & Samanta, D. (2022). A Review on Rural Women's Entrepreneurship Using Machine Learning Models. *International Conference on Innovations in Computational Intelligence and Computer Vision*, 375–395.

Pauly, R., & SpringerLink (Online service). (2021). *Economic instability and stabilization policy: On the path from crises to state directed economies*. Springer Fachmedien Wiesbaden. doi:10.1007/978-3-658-33626-4

Pepe, P. (2022, March). On Lyapunov Methods for Nonlinear Discrete-Time Switching Systems With Dwell-Time Ranges. *IEEE Transactions on Automatic Control*, *67*(3), 1574–1581. doi:10.1109/TAC.2021.3069661

Podder, S. K. (2021). 2022). Impact of Innovative Technology on Quality Education: The Resourceful Intelligence for Smart Society Development. Machine Intelligence and Data Science Applications. *Proceedings of MIDAS*, *132*, 293.

Rimal, Y., Rathore, V. S., Pageni, S., Samanta, D., Karuppiah, M., & Singh, D. (2022). Neural Network based Student Grade Prediction Model. *2022 International Conference on Machine Learning, Computer Systems and Security (MLCSS)*, 22–27. 10.1109/MLCSS57186.2022.00013

Sahu, S. K., Chandra, J., Muloor, K., & Samanta, D. (2023). A Thorough Investigation of Various Goals and Responses for Mobile Software-Defined Networks. In *Designing and Developing Innovative Mobile Applications* (pp. 307–328). IGI Global. doi:10.4018/978-1-6684-8582-8.ch017

Vishnu, J., & ... A strategy for optimal loading pattern of a typical power system — A case study. *2014 International Conference on Advances in Green Energy (ICAGE)*, 112-117. 10.1109/ICAGE.2014.7050152

Yuliia Matvieieva. (2021). *9 The Relationship Between Inequality and Socio-Economic Drivers as a Tool for Sustainable Territorial Development, in Economic Inequality – Trends, Traps and Trade-offs*. River Publishers.

Chapter 15
Influence of Eco-Related Anxiety and Climate Distress on Mental Health

K. Arya
PES University, India

V. Shantha
PES University, India

Zidan Kachhi
https://orcid.org/0000-0002-8317-6356
PES University, India

ABSTRACT

The concept of eco-anxiety, also known as climate distress, refers to a constant dread of environmental disasters brought on by recognizing how inevitably climate change would affect the environment. Eco-anxiety also involves raising concerns for one's own survival and those of future generations. The sense of unease or powerlessness over the impacts of climate variations or changes on the environment and one's own life is an aspect of eco-anxiety. Studying the influence of eco-anxiety, or climate distress, on mental health is crucial because it frequently occurs in the world and has a significant negative impact on people's functioning. Eco-anxiety has positive and negative effects on social behavior, cognition, productivity, and behavioral responses, all in turn, influencing the mental health and well-being of a person. The negative impact of climate change can be reduced by engaging in climate action.

DOI: 10.4018/979-8-3693-2177-5.ch015

INTRODUCTION

The impact of changes or fluctuations in the environment on mental health is increasing. Evidence suggests that increases in temperatures and natural disasters have the ability to influence mental health and can even cause psychological distress. The ever-changing and instability in the environment gives rise to feelings of helplessness, anxiety, depression and anger. The chronic fear or anxiety regarding the changes caused by the environment or ecological damage is referred to as eco-anxiety or climate distress. According to the American Psychological Association, eco-anxiety is defined as the chronic fear of environmental calamities that is caused due to observing the inevitable influence of changes in the climate and the related worry for one's survival.

Climate distress or eco-anxiety involves the feeling of uneasiness or helplessness over fluctuations in the climate on the environment and one's own life. The discomfort could manifest as disturbing ideas or thoughts regarding impending doom or calamities. The uneasiness may also be caused by worrying about humanity and the effects of environmental changes on future generations or one's future and existence. The anxiety or distress produces physiological, emotional and behavioral components. Breathlessness and heart palpitations are caused due to panic rising from changes in one's environment. Other physiological changes could include allergies, and a rise in exposure to diseases. Eco-anxiety causes an increase in stress levels, anxiety, trauma, helplessness, etc. The influence of eco-anxiety on mental health has an influence on the increased use and addiction to substance abuse. Eco-anxiety is not listed in the 'Diagnostic and Statistical Manual of Mental Disorders' (DSM-5), hence, it is not regarded as a disorder or a condition that can be diagnosed. This makes it more important that people are aware of what eco-anxiety is and take steps to reduce its negative consequences.

People do not experience eco-anxiety equally. High stress, depression, feelings of nervousness and guilt, disturbances in sleep and frequent uneasy thoughts among other symptoms are caused due to fear of environmental fluctuations. Eco-anxiety causes one to feel suffocated and breathless or trapped. Among people who face guilt concerning the environment, adults with children tend to possess greater feelings of fear and uneasiness thinking about the future.

According to a systematic review done by Boluda-Verdú, et al. (2022), eco-anxiety is defined as people's reactions of worry, fear, guilt and concern towards climate change threats and degradation of the surrounding environment. The authors included a sample of 12 articles for review. Some authors regard eco-anxiety as a normal response and do not require clinical diagnosis, whereas, others view eco-anxiety as a vital factor in influencing mental health. Eco-anxiety was found to be linked with depression, post-traumatic stress disorder, insomnia, anxiety and stress. Eco-anxiety is also associated with decreasing mental health and an aversion to conceiving children. Eco-anxiety may act as a stressor to mental well-being, particularly in situations where coping strategies are yet to be established.

Coffey et al. (2021), through their systematic review, view eco-anxiety as distress or discomfort caused due to differences in the climate. People undergoing eco-anxiety feel anxious about the future. Climate change causes issues regarding mental health and well-being. 'The Australian Medical Association' stated that change in climate causes harsh consequences to health. The study includes adverse behaviors linked with changes in climate, including irritability, sleeplessness, guilt, frustration and anger. Eco anxiety is often linked to having negative impacts, however, it also has positive consequences which include, hope, self- determination and connection.

Climate change has an impact on the physical and mental health of individuals. Natural calamities and degradation of the environment cause people to develop eco- anxiety as they start fearing the stability of the environment. Awareness about the consequences of fluctuations in the environment has pushed people to develop varying reactions. Some individuals take action and safeguard the environment from more degradation, while others develop anxiety, depression and feelings of guilt.

Padhy et al. (2015), suggests that an increase in temperatures is more likely to increase aggression and rates of violent suicide. Natural disasters, such as, droughts lead to impaired mental health, stress and an increase in suicide of farmers. An increase in disasters with climate change leads to the development of adjustment disorders and depression. As changes in climate and an increase in natural disasters cause more people to migrate in search of employment and shelter, it leads to an increase in acculturation stress. It also causes a rise in physical illnesses which influence negative emotions and psychological distress.

Climate change is found to trigger complicated grief, survivor guilt, fatigue, suicidal tendencies and substance abuse. Climate distress elicits two opposing reactions from people. It negatively affects mental health and increases feelings of hopelessness and despair as reactions towards the problem or fluctuations in the environment. On the contrary, unstable conditions and natural disasters also inspire people to develop compassion, altruistic behavior and an increased sense of meaning and personal growth.

The chapter makes use of a literature review to collect and interpret existing information on the issues of climate distress and eco-anxiety and its influence on mental health. It also aims to understand the impact of eco-anxiety on the emotional, social and behavioral aspects of individuals. The chapter includes information on eco-anxiety and climate distress, its impact on social, emotional, and physiological aspects of individuals and their influence on mental health. Eco-anxiety is a prevalent condition and the chapter covers how it is responsible for influencing one's mental well-being and how it affects an individual's behavior and attitudes.

The primary mission of this chapter is to investigate eco-anxiety and its influence on mental health. It is important to study the impact that eco-anxiety or climate distress has on mental health as it greatly affects the functioning of individuals and is a very common occurrence in the present world. This chapter aims to provide more insight into eco-anxiety and climate distress and understand its effects and implications. The study is valid as eco-anxiety or climate distress is not a much-researched topic and there is a limitation in information on the effects of eco-anxiety and people's response towards climate change and environmental catastrophes. The research gap in the chapter includes the lack of information on ways to help individuals counteract the effects of eco-anxiety and climate distress.

INFLUENCE OF ECO-RELATED ANXIETY ON MENTAL HEALTH

According to a study conducted by Gunasiri, H., et.al. (2022), climate change has adverse effects on health. The study was developed in Australia and the results state that two-thirds of Australians hold the view that the mental health issues caused due to climate change or fear for the future pose the greatest risk factors to health. Most adolescents and young adults turn to substance abuse to distract themselves from stressful situations and alleviate stress caused by fluctuations in the environment. They also feel guilty about how their actions impact the environment (Huizen, J. and Timothy J. 2019).

People who face eco-anxiety or climate distress have a greater chance of suffering from Post-Traumatic Stress Disorder and Generalized Anxiety Disorder. People feel frustrated and worried about the changes in the environment. Many people experience feeling helpless since they are unable to control the consequences of climate change. Adolescents are impacted by the lack of purpose and loss of control which arises from frequent changes in environmental conditions and uncertainty about the environment. They constantly struggle to fit in and adjust to the new circumstances, which contributes to low self-esteem and self-doubt when one is unable to handle the shifting circumstances.

INFLUENCE ON SOCIAL BEHAVIOR

Climate action includes intentional behaviors and participation of individuals in activities that help to decrease the harmful factors in the environment which consists of greenhouse gas emissions or pollution and improve skills required to overcome the adverse consequences of climate change. Young adults are most affected by eco-anxiety, however, they are the most successful group working against the ill- effects of eco-anxiety. Collectively, they aim to improve issues and decrease environmental degradation. The youth take action to address the adverse consequences of anxiety related to the environment and work towards building a more sustainable and healthy future. The sense of belonging to a group that works for a good cause and encourages social inclusion has a positive impact on mental health and overall wellbeing since it promotes the formation of an in-group.

Climate distress and eco-anxiety motivate people to take action. A group of people who have similar goals make an in-group. A person's feeling of social acceptance and validation that comes from belonging to such a group boosts their perception of self-worth. Social inclusion is crucial for mental health because it gives people a sense of significance, worth, and recognition. Individuals who feel socially accepted and a part of groups tend to be less impacted by anxiety related to the environment than those who feel excluded. Some of the concerns and uncertainties about the environment and the future can be reduced by participating in group therapies or just talking about anxiety with other group members. It makes one feel more normal to know that there are other people who experience the same problem.

People's motivation to engage in environmental promotion is one of the positive effects of eco-related anxiety on social behavior. Individuals channel their anxious and uncertain thoughts into constructive deeds and other activities that focus on creating a healthy environment. People who are experiencing climate distress and eco-anxiety make the decision to lead more responsible and ecologically conscious lives, thereby improving their social and personal relationships. These individuals might also take part in educating others in their social circles about sustainable living behaviors.

Though there are numerous positive consequences of eco-anxiety on social behavior and the forming of interpersonal relationships, there also exist a lot of negative effects of eco–anxiety on social behavior. Individuals experiencing high levels of eco-anxiety choose to withdraw from social groups and prefer to stay isolated. They may avoid social gatherings and events due to the recurring feelings of helplessness and loss of control associated with the environment. As eco-anxiety leads to preoccupation with the environment and the changing climatic conditions, it is usually difficult for individuals to fully immerse themselves in social activities or interactions. There may be a high chance of them being mentally distracted, hence, being less focused on maintaining interpersonal relationships. Eco-anxiety also targets mental health, increasing chances of depression and anxiety, which influences social interactions. Some may resort to self-blame and self-doubt for their lack of contributions towards a healthy environment, which affects their confidence and self-esteem, leading to struggles in forming and maintaining relationships. Eco-related anxiety and climate distress may also act as a barrier to forming deep connections with other people who lack the same concern for the environment. Individuals with high eco-anxiety may also resort to criticizing others who they feel are damaging the environment or not taking part in environment-friendly activities, which leads to a negative atmosphere in social settings.

Individuals are influenced differently by eco-anxiety and climate distress. Some are encouraged to work towards a better environment leading to better social behavior. Some may be negatively affected leading to isolation and negative social behavior.

CLIMATE ACTION

Eco-anxiety consists of emotional distress and anxiety in response to the changing environment or environmental threats. It also includes uncertainty regarding the future and worries about the prospective doom of our environment. To counter the feelings of anxiety and worry about the environment, people engage in climate action. Climate action refers to the coordinated efforts made by people, groups, governments or communities to decrease the consequences of changes in the environment and create a more environment-friendly future. The initiatives of climate action aim to tackle the different environmental, economic and social issues brought about by frequent changes in the climate or environmental disasters.

People who engage in climate action learn to control their anxiety regarding uncertainty about the future. They can change negative thoughts into optimistic ones, as they participate in activities to help the environment become healthier. When young people engage in climate action, they feel more confident and at ease with the changing climate and experience less eco-related anxiety. People feel hopeful and optimistic just by knowing they are a part of organizations or initiatives that try to improve the environment. Personal resilience and the capacity to take initiative, enforced by climate action, support mental health and well-being. When confronted with challenges or adverse situations, people's sense of responsibility or ability to take action seemed to favor climate action, contributing to boosting optimism about the future. Supporting and encouraging those with eco-anxiety to engage in climate action can help them build resilience and a sense of control. People who engage in climate action are psychologically and mentally fit as it motivates people to feel resilient, optimistic and enforces a sense of control.

Eco-anxiety can cause positive and negative consequences on social behavior. The positive effects of eco-related anxiety on social behavior include the motivation of individuals to participate in environmental advocacy. People work together to tackle climate change and other related environmental issues. People direct thoughts of worry and uncertainty into positive actions and other activities towards achieving a healthy environment.

BEHAVIORAL CONSEQUENCES OF ECO-RELATED ANXIETY AND CLIMATE DISTRESS

Eco-anxiety and climate distress can influence behavior in different ways. How people make decisions, engage with their surroundings and interact with other individuals.

People who suffer from eco-related anxiety and climate distress choose to behave in environmentally conscious ways and may choose to reduce waste and alter consumption patterns. According to Evans, G.W. (2018), projected behavior is influenced by changes in the climate which includes natural disasters, pollution, elevation or decrease in temperature and frequent changes in weather conditions. Climate-induced distress can motivate individuals and societies to adopt behaviors that help them adapt to the changing environment. Resilience is considered an important trait to have when dealing with the effects of eco-anxiety and climate distress. Resilience refers to the ability to face and overcome hardships or challenges. Individuals who display resilient behaviors are more prepared to cope with eco-anxiety as they have the willpower to cope with the changing environment and overcome the consequences of eco-anxiety. People experiencing eco-anxiety or climate distress may also be more motivated to indulge in environment-friendly activities to reduce their carbon footprint.

Individuals make use of different coping mechanisms to deal with the consequences of eco-anxiety. Some resort to indulging in environment-friendly activities referred to as climate action. Others resort to negative ways such as projecting worry and anger about the environment on others and venting out frustration on them. Individuals experiencing climate distress or eco-anxiety may completely isolate themselves and prefer to stay indoors and avoid going out. This negatively impacts social life and also reinforces feelings of depression and anxiety. They also experience a loss of control start doubting their abilities and develop negative self-concepts.

People experiencing eco-related anxiety often turn to substance abuse to cope with the stress, anxiety and worry about the uncertain future. Some may portray behaviors linked to self-doubt and low self-esteem, as they criticize themselves for the negative state of the environment and not being able to adapt to it. As they fear uncertainty about the environment's future, they are more aversive to risk. They avoid making investments or long-term plans as they do not trust the environment due to its constant changes. Such people are excessively careful, and worry about the smallest fluctuations in the climate. Additionally, they make an effort to keep themselves informed about environmental changes by reading up on ecological trends and related research.

NATURAL DISASTERS

According to Huizen, J. and Timothy J. (2019), anxiety regarding the environment can occur from experiencing or having loved ones at high risk of extreme weather conditions linked to changes in the climate, resulting in natural disasters. Fear of natural disasters is associated with eco-related anxiety or climate distress. As natural disasters are usually unpredictable, though scales that can predict natural disasters have been invented, they cause a great deal of worry to people. Natural disasters are environmental events that are considered unfavorable and are brought on by the earth's natural processes. They are disastrous and frequently result in death, loss and destruction. Natural disasters pose a huge threat to several people, disrupt services and other networks, deplete community resources and adversely affect the mental and physical well-being of individuals.

Natural disasters can have significant effects on mental health, both short-term and long-term. The influence natural disasters have on mental health can vary from person to person, but there are few common psychological responses to natural disasters (Gargano, S.P. and Saeed, A. 2022). Post-traumatic stress disorder is common among survivors of natural disasters. Post-traumatic stress disorder (PTSD) is a mental health condition that develops in people who have witnessed or been involved in a stressful or traumatic event. Post-traumatic stress disorder can affect people of any age and present itself in different ways. It is usually triggered by exposure to a stressful situation which reminds the person of the traumatic event. People who are diagnosed with post-traumatic stress disorder experience flashbacks, severe anxiety, panic attacks, nightmares and hypervigilance. As natural disasters cause too much destruction and people targeted by natural disasters often face trauma, loss and grief, they are usually affected by post-traumatic stress disorder. News about natural disasters or even slight similarities to the same environment of the natural disasters can trigger panic attacks, anxiety, flashbacks, etc.

Natural disasters cause survivors to develop acute stress reactions, which are psychological responses that occur immediately after the occurrence of a traumatic event (Classen, C., et.al. 1995). Acute stress reactions or disorders are similar to post-traumatic stress disorder, but acute stress disorders produce a shorter duration of responses and are generally less intense when compared to post-traumatic stress disorders. Individuals with 'acute stress disorders' experience dissociation and may feel a sense of detachment or feel as if the events happening around them are figments of their imagination. They also experience intrusive thoughts, distressing flashbacks, and recurrent memories regarding the trauma inducing event. They also experience irritability, insomnia, and severe anxiety. Anxiety disorders are also popular among those experiencing eco-anxiety due to natural disasters. Survivor's guilt may develop among individuals who managed to live through the natural disaster. They begin to accept responsibility for the event that happened and feel guilty for not being able to avert the natural disaster and spare other people's lives. Physical health can be affected by natural disasters which also affect mental well-being.

Natural disasters lead to the destruction of resources and cause harm to many people. Natural disasters cause displacement and force people to migrate due to loss of livelihood, homes and security. The people who live in areas constantly targeted by natural disasters, such as, along the coast, face eco-anxiety or climate distress and live in constant fear of the occurrence of a natural disaster. Displacement caused by natural disasters has a significant long-lasting influence on the individual's well-being. The sense of being forced out of one's home and society, also accompanied by loss and grief can negatively affect mental health. They also experience constant fear of losing control and uncertainty. Some of the people troubled by natural disasters have to migrate to other places, and they experience acculturation stress. Trying to fit in with another community and following a different culture can adversely impact mental health.

EFFECTS OF ECO-ANXIETY ON PRODUCTIVITY AND MOOD

Clayton and Karazsia's (2020) study on how the changes in the climate promote anxiety, demonstrates the multidimensional aspect of climate distress. Four distinct dimensions were determined, they included environmentally favorable behavior, experience of environment and climate change, cognitive and emotional problems and limitation in productivity. The cognitive and emotional aspects seemed to be the most troubled due to climate distress. As a consequence of cognitive and emotional imbalance, daily activities were affected. Productivity and efficiency decreased and people remained more distracted due to worries and fear of uncertainty arising from eco-anxiety.

Extreme weather events or climatic conditions lead to frustration and build-up of anger and irritation. People who live in areas vulnerable to natural disasters constantly witness loss of biodiversity and other disastrous effects of extreme weather conditions. They have a constant low mood, and they frequently experience worry and anxiety, more so than other people. Studies also show that a rise in temperatures causes people to feel angry and frustrated. People are more aggressive and irritated when the weather is extremely hot. Aggressive behavior is more likely seen when temperatures increase. A study states that an 18 per cent increase in aggressive behavior and violence between prisoners was found on days when the temperature was high (Christensen, J. 2023). High temperatures make people feel uncomfortable and increase unfavorable emotions.

Climate change also causes worry about food production and security. Farmers and those involved with agriculture and the production of food, are vulnerable to eco-anxiety. Communities that rely on nature for livelihoods such as hunters, fishermen and food gatherers are also affected by eco-anxiety. They live in fear of the changes in the climate that can destroy their source of livelihood and negatively impact their lives.

Anxiety related to environmental concerns has a significant impact on people's moods. The emotional distress and fear linked to eco-anxiety or climate distress lead to mood- related changes. These changes differ from person to person depending on the intensity of uncertainty or anxiety faced by them. The persistent and recurrent anxiety and concern regarding the ecological challenges and climate change produce feelings of unease, restlessness and stress, affecting one's mood.

Climate distress impacts productivity and the emotional and psychological burden of eco-anxiety adversely affects decision-making, concentration and engagement in tasks. Climate distress or eco-anxiety reduces concentration due to preoccupation with environmental issues and a sense of worry regarding general existence. Individuals with eco-anxiety face challenges both in their personal and professional lives. Anxiety and climate distress results in procrastination and people may avoid taking decisions due to their fear of the uncertain future. Increased stress levels negatively affect productivity and it inhibits individuals from being able to consider multiple viewpoints. Differences in beliefs or values regarding the environment may also lead to conflict among individuals in the workplace. This affects interpersonal relationships and due to these differences, they find it hard to work with their other colleagues. Increased stress and anxiety lead to burnout and reduced levels of resilience.

Despite the negative effects that have been correlated to eco-anxiety, there can be no denying that there are some positive outcomes to eco-anxiety too. Eco-anxiety acts as a catalyst for people to be more aware of their environment and the consequences of their actions on it. It prompts people to do more research on themes such as climate change, pollution, etc. There have been observations that people start environment awareness projects and take the necessary steps to prevent carbon footprint etc. as a result of eco-anxiety. As research states, like-minded people gravitate to form a community. Hence, people who are affected by sci-anxiety are likely to form a community to spread awareness through marches, conducting awareness workshops or lectures, etc. This makes people more wary and prompts them to think of preventative measures for environmental issues and the possibility of it becoming a bigger problem. One would assume, since issues such as global warming have been ever so present for many years that eco-anxiety would also have been a prevalent problem for quiet many years but research shows that it's more of a recent psychological phenomenon. Having anxiety can be extremely overwhelming and can bring out a feeling of helplessness. People tend to feel that they are too minuscule to make important changes when compared to the sheer size of current environmental problems and the possibility of future environmental conditions. This in turn can make one feel isolated with anxiety, and have physical effects such as headaches, stomachaches, etc. (Coffey et al., 2021)

A poll conducted by Friends of Earth in 2020 states that 70 per cent of people within the age group of 18 to 24, are comparatively more worried about climate and environmental changes with their impact this year when compared to last year in the UK. Our emotional response to our knowledge that the root cause of an environmental issue is us along with the knowledge that it is a struggle to keep the damages as minimal as possible worsens our stress and anxiety.

Children and young adults are at a higher risk than adults of being affected by ecological changes. Since, we are at an age where there is information overload through news channels, articles, YouTube videos, etc, this can likely cause distress, necessary and unnecessary to an extent. Parents need to supervise the type of content that kids are consuming. Spreading awareness to parents on how to deal with such eco-anxiety in children is essential. This feeling of powerlessness can be overwhelming for a child and make them feel dissociated. It is difficult for them to navigate and understand such feelings, at this time it's crucial for them to address it and for someone to help them process such emotions and make them feel validated to be feeling the way that they are feeling. Acknowledging these feelings will prevent them from avoiding the shame or guilt that may accompany. Although there are negative effects and connotations attached to the term eco-anxiety and there are people who do not consider this as a technical mental health issue, we need to understand that the severity of it is subjective and there are actual cases where this can hamper the day to day lives of few individuals. It is vital for us to note that those who are affected by eco-anxiety indicate that people care and worry about their environment and this way of looking at it might help people in their perspective. Rather than have strict labels, it is better to look at it as a spectrum. (Hickman, 2020).

A mild form of eco-anxiety would be sometimes worrying about the environment. But these people are most likely to get over these feelings of worry with some reassurance that "everything will be okay" and tend to think there are other people who are more capable of finding solutions to the environmental problems causing them to worry. These people are more optimistic in nature and are okay with relying on others to solve problems without themselves being the ones to come up with a solution. A more moderate level of eco-anxiety would be people often worrying about issues like climate change, global warming, etc. These people tend to rely upon and believe that only "the experts" can find. It's more difficult to reduce the distress caused to these people. As we have noticed, such people tend to get conscious and make decisions like going completely vegan switching to a vegetarian diet, or deciding and choosing to not use products such as makeup that is tested on animals, using only cruelty-free products, etc. If this worry extends to a point that a person is not able to function normally in their lives then it can be categorized into severe side of the spectrum.

In extreme cases, every day, eco-anxiety may make individuals feel quite disturbed and concerned. They could be concerned about societal breakdown and climate change. Their anxiousness may be difficult to control, and they may feel guilty or humiliated. They may have little confidence in others to identify or act on climate-change solutions. They may be willing to cut of relationships with persons who are denying the climate crisis. They may choose not to have children or quit flying, among other things. They may endure significant cognitive changes, such as intrusive thoughts and difficulty sleeping. They may be terrified and have little faith in anyone to aid them. They could be unable to control their emotions, think clearly, or focus. For example, a person going about their normal day can get triggered when he/she sees other people using plastic cups or throwing wrappers around, and all of this can have an extremely bad effect on that

COVID gave us a clear example of how a lot of our properties and activities such as fights, smoke, and industrial corporations affect the environment. Since there was a lockdown, less human activity, and less pollution, there were many news outlets that covered wildlife species roaming freely on roads, etc. This alone is a great example of how the elimination of a few human activities which cause harm to the ecosystem and environment can have drastic positive effects on our surroundings. Commonly this anxiety is treated through medication, therapy, etc. Exercise is one of the practices that has been noted to help people, as doing physical activities releases certain chemicals in the brain such as endorphins.

Children find it more difficult to feel reassured or find comfort in thoughts that there are other people who may find solutions to the problems, they take it more as a personal responsibility and hence they feel more distressed which can get as extreme as having suicidal thoughts because of extreme feelings of helplessness. There needs to be a maintained balance of the world inside us that runs through thoughts, feelings, imaginations, etc. and the world around us which is our reality. We cannot just concentrate on the outside world without acknowledging and understanding us humans, that we are flawed, we are powerful only to an extent, we have feelings, etc. At the same time, we cannot just focus on the world inside us and ignore the outside reality. This is one way that we can become very ignorant and lose concern for our planet, its future, etc (*The Oxford Handbook of Acceptance and Commitment Therapy*, 2023).

To an extent we need to come to the realization that at any point in time, there is a possibility of the world ending someday, global warming becoming severe, etc. We cannot ever be at a point in time where we can be assured with a hundred percent security that the state of our environment will be healthy. Some ways in which a person can be affected by natural calamities are; they have personally lived through such experiences and as an affected effect can have eco-anxiety, PTSD, etc., seeing or hearing about their loved one's having gone through such experiences and hence having stress as a result, and finally, seeing news or just getting lost in the thought spiral of the possibilities of natural calamities, etc.

Different age groups have anxiety in varied aspects too. Let's take an example of adults who are parents may have eco-anxiety due to the fear of the effects of such calamities on their children's future in terms of physical and mental health. These days quite a few people choose to not have kids because they do not want to put them through these experiences which can be scary. Personalities also come into play about eco-anxiety, people who are more predisposed to being sensitive to changes in the environment are comparatively more likely to be affected by calamities. We also need to be mindful while consuming information through media outlets as there are many chances of them exaggerating the news, which might trigger unnecessary anxiety among people. Some habits that might help are learning to ground yourself, developing daily healthy routines, being aware of your emotional state, having control over the amount of media and news being consumed, etc. Although reason is important we should keep in mind that emotions are equally as important and we need to develop our emotional literacy, become aware of the emotions we are currently feeling, be able to articulate them and possibly label them, etc. are all important (Pihkala, 2020).

Among some of the coping strategies, de-emphasizing the threat is commonly used. Minimizing the extent of the threat and its importance, thinking that it is not your responsibility to deal with it, relying on better others to handle the problem and come up with appropriate solutions, etc. are some examples. Another strategy that people use which is not necessarily a healthy one is distraction and isolating the negative emotions without properly acknowledging it. This can pile up and have long-term impacts on a person's mental health. It's a minority of people who choose to share such distress and actively seek help for it although the numbers are getting better as the years go by and there is more awareness regarding the benefits of seeking professional help. One healthy coping strategy that needs to be mentioned is when people analyze the problem, take a reality check, think of possible solutions, and take the necessary concrete measures to work towards attaining the solution. For example, choosing to not eat meat, using only biodegradable packaged products, etc. (*ECO-ANXIETY on JSTOR*, n.d.).

Forming a community and working towards the same goal together has proven to help people feel this belongingness when compared to working alone. Adults need to take more measures to educate and spread awareness to reduce the effects of eco-anxiety on the younger generation. More than younger children, older teenagers and people in their early adulthood are the ones pessimistic and in distress in terms of the future of our ecosystem and environment. Taking all this into consideration, intergenerational understanding and support is necessary. It can reduce the overwhelming feelings that the youth have and help them even if it's in minute steps. This pressure to take action immediately and get rid of the existing problem can be pressuring but before that critical thinking about the problem, discussions considering different points of view, and an open discussion are important too. Occasional thought or worry is considered normal and even good as this shows awareness but habitual anxiety or worry is a major indicator of needing some professional help and care.

There are multiple issues related to mental health and ecological states. Being aware of these types and what they mean in turn can help us understand ourselves better which in turn makes the journey of healing, and attaining a healthy mental state much easier. When a person feels incapable of worrying about the ecological state of the world, it is termed eco-paralysis. Eco-guilt is when a person feels guilty with the knowledge that any of their actions and behaviors have to some degree added to the worsening environmental state. Eco-grief is a phenomenon where a person has this intense sense of sadness over some sort of ecological loss, be it the destruction of a particular forest, the extinction of a particular species of animals, birds, or insects or it could also be worsening of pollution around them. It could also be that some calamity has taken place in the past and a person feels a certain level of guilt thinking that they could have done something and contributed in some way at least to either make it better or prevented it from happening to a certain extent (Ágoston et al., 2022).

Problem-focused coping techniques attempt to eliminate a stressor. Problem-focused coping occurs when a person is directly focusing on the stressor and working on it to be removed. Planning, problem-solving, or eliminating the stressor are all different ways of problem-focused coping. Emotion-focused coping is attempting to cope with your emotional reaction to the stressor. Emotion-focused coping is utilized when you strive to diminish, eliminate, or merely endure your emotional response to a stressor. Withdrawal, expressing anger and frustration, seeking emotional support, diversions, rumination, and resignation acceptance (accepting that the situation will always remain) are some examples. Research shows that problem-focused techniques are the most commonly suggested strategy to deal with eco-anxiety and it is seen to help people and help remove the stressor or at least help deal with the stressor without being overwhelmed with emotions.

Vulnerable populations are more susceptible to such distress because they might be one of the groups to take the hits of a natural calamity the hardest. This distress is rooted in past experiences, fear, etc. and what fuels these worries is uncertainty. The guilt is higher for people whose area or country's ecological footprint is above average when it is objectively compared to other countries.

This can elicit two kinds of reactions from people, either this motivates them to be pro-change and action, taking necessary steps that are needed. Another type of reaction could be losing the motivation to do anything to change the trajectory of the distress. A person's perspective can play a major factor in their reaction to such anxiety and stress (Ágoston et al., 2022).

CONCLUSION

The effects of climate changes like global warming etc. are increasing especially in recent years. The rise in temperatures in the environment, natural disasters such as tsunamis and floods, and degradation of the environment can induce some amount of psychological distress and eco-anxiety, which in turn can lead to a range of negative emotions and related behaviors. Eco-anxiety is a relatively new term that describes the chronic fear or anxiety related to environmental changes or ecological damage. It is not recognized officially as a mental disorder in the DSM-5, but it is still very real and is a significant enough issue that has an effect on many people. Eco-anxiety can show itself in different ways, there may be some physiological, emotional, and also behavioral symptoms. Some people who are more aware and care more about the state of their environment and its protection are comparatively more likely to experience and suffer from eco-anxiety. Sudden or gradual changes in climate and extreme weather conditions can trigger feelings of complicated grief, survivor guilt, fatigue, suicidal tendencies, substance abuse and much more. As with many things in the world, eco-anxiety can cause both positive and negative behaviors, such as hope, connection, empowerment, irritability, sleeplessness, guilt, frustration, and anger. Young people are especially affected by eco-anxiety, as they are more likely to be aware of environmental issues and the potential impact of their present activities on their future. However, his young generation is also the most significant group in terms of working towards and bringing about positive changes to the environment. Being part of a group that works towards a good cause has positive effects on mental health and well-being as it induces the feeling of social inclusion and creation of in-group. It is essential for us to recognize the consequences of climate change on mental health and well-being and take the necessary steps to reduce its negative consequences. This includes spreading awareness of eco-anxiety, providing support and resources for those affected, and taking action by addressing the root causes of climate change. By working together, we can make a healthier and more sustainable future for ourselves and future generations. Furthermore, the impact of changes in climate on mental health and well-being is not particularly limited to eco-anxiety. Natural calamities, such as hurricanes, floods, and wildfires, can be the cause of trauma, post-traumatic stress disorder (PTSD), and depression. Displacement and loss of homes, in turn, communities, and loved ones can lead to severe grief, anxiety, and depression. Climate change can also worsen existing mental health conditions, such as anxiety and depression, by snowballing stressors and reducing access to resources. Marginalized communities, such as low-income and indigenous populations, are excessively affected by the mental health impacts of climate change. They often have limited access to resources and are more vulnerable to the effects of natural disasters and environmental degradation. Climate justice, which aims to address the inadequate distribution of the impacts of changes in the climate, is therefore crucial in promoting mental health and prioritizing well-being. We can conclude that the impact of climate change on mental health and well-being is a complex and multifaceted issue that requires attention and action. By identifying the mental health impacts of climate change, promoting awareness, providing support and resources, and taking action to address the root causes, we can create a healthier and more sustainable future for ourselves and future generations.

REFERENCES

Ágoston, C., Urbán, R., Nagy, B., Csaba, B., Kővári, Z., Kovács, K., Varga, A., Dúll, A., Mónus, F., Shaw, C. A., & Demetrovics, Z. (2022, January 1). *The psychological consequences of the ecological crisis: Three new questionnaires to assess eco-anxiety, eco-guilt, and ecological grief.* Climate Risk Management; Elsevier BV. doi:10.1016/j.crm.2022.100441

Boluda-Verdú, I., Senent-Valero, M., Casas-Escolano, M., Matijasevich, A., & Pastor-Valero, M. (2022, December 1). *Fear for the future: Eco-anxiety and health implications, a systematic review.* Journal of Environmental Psychology; Elsevier BV. doi:10.1016/j.jenvp.2022.101904

Christensen, J. (2023). *Hot under the collar? Heat can make you angry and even aggressive, research finds.* https://edition.cnn.com/2023/09/05/health/heat-anger-wellness/index.html

Cianconi, P., Betrò, S., & Janiri, L. (2020). The Impact of Climate Change on Mental Health: A Systematic Descriptive Review. *Frontiers in Psychiatry*, *11*, 74. Advance online publication. doi:10.3389/fpsyt.2020.00074 PMID:32210846

Clayton, S., & Karazsia, B. (2020). Development and validation of a measure of climate change anxiety. *Journal of Environmental Psychology*, *69*, 101434. doi:10.1016/j.jenvp.2020.101434

Coffey, Y., Bhullar, N., Durkin, J., Islam, M. S., & Usher, K. (2021, August 1). *Understanding Eco-anxiety: A Systematic Scoping Review of Current Literature and Identified Knowledge Gaps*. The Journal of Climate Change and Health; Elsevier BV. doi:10.1016/j.joclim.2021.100047

Coffey, Y., Bhullar, N., Durkin, J., Islam, M. S., & Usher, K. (2021, August). Understanding Eco-anxiety: A Systematic Scoping Review of Current Literature and Identified Knowledge Gaps. *The Journal of Climate Change and Health*, *3*, 100047. doi:10.1016/j.joclim.2021.100047

ECO-ANXIETY on JSTOR. (n.d.). https://doi.org/ doi:10.2307/26798430

Effects of climate change on future generations. (n.d.). Save the Children. https://www.savethechildren.org/us/what-we-do/emergency-response/climate-change#:~:text=Extreme%20temperatures%20leave%20many%20families,sensitive%20to%20disease%20and%20pollution

Evans, G. W. (2019). Projected behavioural impacts of global climate change. *Annual Review of Psychology*, *70*(1), 449–474. doi:10.1146/annurev-psych-010418-103023 PMID:29975596

Grose, A. (2020, June 9). *A Guide to Eco-Anxiety*. Watkins Publishing.

Gunasiri, H., Wang, Y., Watkins, E., Capetola, T., Henderson-Wilson, C., & Patrick, R. (2022). Hope, Coping and Eco-Anxiety: Young People's Mental Health in a Climate-Impacted Australia. *International Journal of Environmental Research and Public Health*, *19*(9), 5528. doi:10.3390/ijerph19095528 PMID:35564923

Hickman, C. (2020, October 1). We need to (find a way to) talk about . . . Eco-anxiety. *Journal of Social Work Practice*, *34*(4), 411–424. doi:10.1080/02650533.2020.1844166

Hogg, T. L., Stanley, S. K., O'Brien, L. V., Wilson, M. S., & Watsford, C. R. (2021). The Hogg Eco-Anxiety Scale: Development and validation of a multidimensional scale. *Global Environmental Change*, *71*(102391), 1–10. doi:10.1016/j.gloenvcha.2021.102391

Howatt, B. (2020, February 12). How does climate-driven eco-anxiety affect our performance at work? *The Globe and Mail*. https://www.theglobeandmail.com/business/careers/workplace-award/article-how-does-climate-driven-eco-anxiety-affect-our-performance-at-work/

Identifying Types of Eco-Anxiety, Eco-Guilt, Eco-Grief, and Eco-Coping in a Climate-Sensitive Population: A Qualitative Study. (n.d.). International Journal of Environmental Research and Public Health; Multidisciplinary Digital Publishing Institute. doi:10.3390/ijerph19042461

Koopman, C., Classen, C., Cardeña, E., & Spiegel, D. (1995, January). When disaster strikes, acute stress disorder may follow. *Journal of Traumatic Stress*, *8*(1), 29–46. doi:10.1002/jts.2490080103 PMID:7712057

Legg, T. J., & Raypole, C. (2020). *Climate Change Taking a Toll on Your Mental Health?* https://www.healthline.com/health/eco-anxiety

Mathers-Jones, J., & Todd, J. (2023). Ecological anxiety and pro-environmental behaviour: The role of attention. *Journal of Anxiety Disorders*, *98*, 102745. doi:10.1016/j.janxdis.2023.102745 PMID:37480627

Padhy, S. K., Sarkar, S., Panigrahi, M., & Paul, S. (2015, January 1). *Mental health effects of climate change*. Indian Journal of Occupational and Environmental Medicine; Medknow. doi:10.4103/0019-5278.156997

Pihkala, P. P. (2020, December 4). *Eco-Anxiety and Environmental Education*. Sustainability; Multidisciplinary Digital Publishing Institute. doi:10.3390/su122310149

Talukder, B., Van Loon, G. W., Hipel, K. W., Chiotha, S., & Orbinski, J. (2021). Health impacts of climate change on smallholder farmers. *One Health*, *13*, 100258. doi:10.1016/j.onehlt.2021.100258 PMID:34027006

The Oxford Handbook of Acceptance and Commitment Therapy. (2023, June 20). Oxford University Press.

Compilation of References

Aadhar, S., & Mishra, V. (2019, November 6). A Sustainable rise in the area and population affected by dryness in South Asia under 1.5 Degree Celsius and 2.5 Degree Celsius Warmer Worlds, Environmental Research Letter. *IOP Science, 14*(11). Advance online publication. doi:10.1088/1748-9326/ab4862

Abdul Khalil, H. P. S., Bhat, I. U. H., Jawaid, M., Zaidon, A., Hermawan, D., & Hadi, Y. S. (2012). Bamboo fibre reinforced biocomposites: A Review. *Materials &amp. Materials & Design, 42*, 353–368. doi:10.1016/j.matdes.2012.06.015

Aberdour, M. (2007). Achieving software quality in open-source software. *IEEE Sustainable Software Testing, 24*(1). 58-64.

Adams, M. (2021). *Climate Change and Mental Health: A Comprehensive Analysis*. Academic Press.

Adhikari, S. P., Meng, S., Wu, Y.-J., Mao, Y.-P., Ye, R.-X., Wang, Q.-Z., Sun, C., Sylvia, S., Rozelle, S., Raat, H., & Zhou, H. (2020). Epidemiology, causes, clinical manifestation and diagnosis, prevention and control of coronavirus disease (COVID-19) during the early outbreak period: A scoping review. *Infectious Diseases of Poverty, 17*(9), 29. doi:10.1186/s40249-020-00646-x PMID:32183901

African Union. (1969). *Convention Governing Specific Aspects of Refugee Problems in Africa.* Author.

Afrin, T., Tsuzuki, T., Kanwar, R. K., & Wang, X. (2012). The origin of the antibacterial property of bamboo. *Journal of the Textile Institute, 103*(8), 844–849. doi:10.1080/00405000.2011.614742

Afzali, R., GharehBeygi, M., & Yazdanpanah Dero, Q. (2020). Climate changes and food policies: Economic pathology. *Climate Risk Management, 30*, 100249. doi:10.1016/j.crm.2020.100249

Agarwal, Nath, & Chowdhury. (2012). Sustainable approaches and good practices in green software engineering. *International Journal of Research and Reviews in Computer Science, 3*(1).

Ágoston, C., Urbán, R., Nagy, B., Csaba, B., Kővári, Z., Kovács, K., Varga, A., Dúll, A., Mónus, F., Shaw, C. A., & Demetrovics, Z. (2022, January 1). *The psychological consequences of the ecological crisis: Three new questionnaires to assess eco-anxiety, eco-guilt, and ecological grief.* Climate Risk Management; Elsevier BV. doi:10.1016/j.crm.2022.100441

Ahmad, Z., Maskat, R., & Mohamed, A. (2023). Harnessing Natural Language Processing for Mental Health Detection in Malay Text: A Review. *2023 4th International Conference on Artificial Intelligence and Data Sciences (AiDAS)*, 29-35. 10.1109/AiDAS60501.2023.10284653

Ai, H., Modi, P., & Mishra, V. (2019, September). Increased Flood Risk In Indian Sub-Continent Under the Warming Climate, Weather and Climate Extremes. *Weather and Climate Extremes, 25*, 100212. doi:10.1016/j.wace.2019.100212

Ait-Yahia Ghidouche, K., & Ghidouche, F. (2019). Community-based ecotourism for preventing overtourism and tourism phobia: Algerian associations' viewpoints. *Worldwide Hospitality and Tourism Themes, 11*(5), 516–531. doi:10.1108/WHATT-06-2019-0035

Ajith, M., & Calhoun, V. D. (2023). Functional Network Connectivity Based Mental Health Category Prediction from Rest-fMRI Data. *2023 IEEE 20th International Symposium on Biomedical Imaging (ISBI)*, 1-5. 10.1109/ISBI53787.2023.10230721

Akbar, D. A., Bewasana, D. A. D., Mahfud, A. Z., & Yulita, T. (2023). InnerCheck: A New Breakthrough for Early Detection of Mental Health. *2023 IEEE International Conference on Cryptography, Informatics, and Cybersecurity (ICoCICs)*, 281-286. 10.1109/ICoCICs58778.2023.10276558

Alam, M. S., & Aktar, H. (2021). The Effect of Social Media on Student Academic Performance: A Case Study at the Islamic University of Bangladesh. *International Journal on Transformations of Media, Journalism & Mass Communication.*

Ali, A. (2018). issue of income inequality under the perceptive of macroeconomic instability: An empirical analysis of pakistan. *Pakistan Economic and Social Review, 56*(1), 121–155.

All India Monthly rainfall Series. (n.d.). https://tinyurl.com/r9bfuzfu

Allgood, L., & McNamara, K. E. (2017). Climate-induced migration: Exploring local perspectives in Kiribati. *Singapore Journal of Tropical Geography, 38*(3), 370–385. doi:10.1111/sjtg.12202

Almaleh, A. (2023). Machine Learning-Based Forecasting of Mental Health Issues Among Employees in the Workplace. *2023 IEEE International Conference on Industry 4.0, Artificial Intelligence, and Communications Technology (IAICT)*, 118-124. 10.1109/IAICT59002.2023.10205620

Alnaser, A. S., & Al-Shibly, M. S. (2020). *Impacts of social media and demographical characteristics on university admissions case of Jordanian private universities.* Palarch's Journal of Archeology of Egypt.

Al-Rahmi, W. M., & Othman, M. S. (2013). The impact of social media use on academic performance among university students: A pilot study. *Journal of Information Systems Research and Innovation.*

Alvi, A. K., Kayani, U. S., & Lakhan, A. B. (2022). *Relationship of social media addiction with academic performance;mediating role of student engagement.* Gomal University Journal of Research. doi:10.51380/gujr-38-02-03

Amin, Z., Mansoor, A., & Hussain, S. R. (2016). Impact of Social Media of Student's Academic Performance. *International Journal of Business and Management Invention* .

Anand, N., Pant, L. M., Alam, T., Pundir, S., Thomas, L., & Rakshith, U. R. (2023). Artificial Intelligence's Contribution To Mental Health Education. *2023 International Conference on Sustainable Computing and Data Communication Systems (ICSCDS),* 462-467. 10.1109/ICSCDS56580.2023.10104905

Anastasiadou, C., & Vettese, S. (2018). Digital revolution or plastic gimmick? Authenticity in 3D souvenirs. In Tourism Social Science Series (Vol. 24). doi:10.1108/S1571-504320180000024011

Anderson, C. A. (2001). Heat and Violence. *Current Directions in Psychological Science*, *10*(1), 33–38. doi:10.1111/1467-8721.00109

Anker, D. (2001). Refugee status and violence against women in the domestic sphere: The non-state actor question. *Georgetown Immigration Law Journal, 15*(3), 391–402.

Archila, H., Kaminski, S., Trujillo, D., Zea Escamilla, E., & Harries, K. A. (2018). Bamboo reinforced concrete: A critical review. *Materials and Structures*, *51*(4), 102. Advance online publication. doi:10.1617/s11527-018-1228-6

Ardenne, P. (2021, June 4). From Eco-anxiety to Resilience. *Critique D'art*, *56*(56), 47–59. doi:10.4000/critiquedart.76145

Ardito, L., Procaccianti, G., Torchiano, M., & Vetrò, A. (2015). Understanding green software development: A conceptual framework. *IT Professional*, *17*(1), 44–50. doi:10.1109/MITP.2015.16

Ari, T., & Gokpinar, F. B. (2020). Climate-Migration: A Security Analysis within the Context of Green Theory. *Uluslararası İlişkiler. International Relations*, *17*(68), 41–53. https://www.jstor.org/stable/26980735

Askland. (2022). Beyond Migration: A Critical Review of Climate Change Induced Displacement. *Environmental Sociology, 8*(3), 267–278. . doi:10.1080/23251042.2022.2042888

Asmundson, G. J. G., & Taylor, S. (2020a). Coronaphobia: Fear and the 2019-nCoV outbreak. *Journal of Anxiety Disorders*, 70. PMID:32078967

Asmundson, G. J. G., & Taylor, S. (2020b). How health anxiety influences responses to viral outbreaks like COVID-19: What all decision-makers, health authorities, and health care professionals need to know. *Journal of Anxiety Disorders*, *71*, 102211. doi:10.1016/j.janxdis.2020.102211 PMID:32179380

Bai, L., & Han, X. (2023). Design of A Mental Health Assessment System Based on Fuzzy Evaluation. *2023 International Conference on Data Science and Network Security (ICDSNS),* 1-6. 10.1109/ICDSNS58469.2023.10245787

Bajaj, V., Bathija, R., Megnani, C., Sawara, J., & Ansari, N. (2023). Non-Invasive Mental Health Prediction using Machine Learning: An Exploration of Algorithms and Accuracy. *2023 7th International Conference on Intelligent Computing and Control Systems (ICICCS),* 313-321. 10.1109/ICICCS56967.2023.10142504

Baker, L., & Smith, R. (Eds.). (2019). *Ethical Considerations in Environmental Psychology*. Routledge.

Balk, S. (2019). *Deciding Whether to Migrate and Where to Go.* https://disasterdisplacement.org/wp-content/uploads/2019/04/dc_2019-04.pdf

Banerjee, D. (2020). The COVID-19 outbreak: Crucial role the psychiatrists can play. *Asian Journal of Psychiatry*, *50*, 51. doi:10.1016/j.ajp.2020.102014 PMID:32240958

Bao, Y., Sun, Y., Meng, S., Shi, J., & Lu, L. (2020). 2019-nCoV epidemic: Address mental health care to empower society. *Lancet*, *22*(395), e37–e38. doi:10.1016/S0140-6736(20)30309-3 PMID:32043982

Bardsley, D. K., & Hugo, G. J. (2010). Migration and climate change: Examining thresholds of change to guide effective adaptation decision-making. *Population and Environment*, *32*(2-3), 238–262. doi:10.1007/s11111-010-0126-9

Baudon, P. (2021, May). A scoping review of interventions for the treatment of eco-anxiety. In Analysis, 5(1), 82–85. doi:10.1016/j.inan.2021.02.005

Becken, S. (2013). A review of tourism and climate change as an evolving knowledge domain. *Tourism Management Perspectives*, *6*, 53–62. doi:10.1016/j.tmp.2012.11.006

Behera, P., & Balaji, S. (2021, June 1). Health benefits of fermented bamboo shoots: The twenty-first century green gold of Northeast India. *Applied Biochemistry and Biotechnology*, *193*(6), 1800–1812. doi:10.1007/s12010-021-03506-y PMID:33496924

Ben Mohamed, A., Van Duivenbooden, N., & Abdoussallam, S. (2002). Impact of climate change on agricultural production in the Sahel–Part 1. Methodological approach and case study for millet in Niger. *Climatic Change*, *54*(3), 327–348. doi:10.1023/A:1016189605188

Benton, A. (2015). *Priority species of bamboo*. doi:10.1007/978-3-319-14133-6_2

Berry, H. L., Bowen, K., & Kjellstrom, T. (2010). Climate change and mental health: A causal pathways framework. *International Journal of Public Health*, *55*(2), 123–132. doi:10.1007/s00038-009-0112-0 PMID:20033251

Berry, H. L., Waite, T. D., Dear, K. B., Capon, A. G., & Murray, V. (2018). Human health and climate change: Leverage points for adaptation in urban environments. *International Journal of Environmental Research and Public Health*, *15*(6), 1232. PMID:29891795

Bhambri, P., & Nischal, P. (2013). *Emergence of the Knowledge Economy and Knowledge Management and their Importance in the Talent Acquisition Front*. Paper presented at the Seminar on NAAC Sponsored National Seminar on Quality Culture in Education.

Bhambri, P., & Gupta, O. P. (2013). A Survey of HPC Applications Appropriate for Execution on DSM and Grid System. *PIMT Journal of Research*, *5*(2), 83–87.

Bhattacharya, Thomas, Soni, Roy, Sarangi, & Kanawade. (2023). Opposite Trends in Heat Waves and Cold Waves Over India. Journal of Earth System Science, (67).

Bhavani, K., & Meeramani, N. (2023). Social media technology and its impact on students' learning skills. In R. Suchithra (Ed.), *International conference on Recent Developments in Science and Technology, Engineering and Management*. Academic Press.

Biermann, F., & Boas, I. (2008). Protecting climate refugees: The case for a global protocol. *Environment, 50*(6), 8–17. doi:10.3200/ENVT.50.6.8-17

BioOne VISTA. (n.d.). *Climate Change*. https://tinyurl.com/988as3u7

Böck, F. (2014). Green gold of Africa–Can growing native Bamboo in Ethiopia become a commercially viable business? *Forestry Chronicle, 90*(5), 628–635. doi:10.5558/tfc2014-127

Boluda-Verdú, I., Senent-Valero, M., Casas-Escolano, M., Matijasevich, A., & Pastor-Valero, M. (2022, December 1). *Fear for the future: Eco-anxiety and health implications, a systematic review*. Journal of Environmental Psychology; Elsevier BV. doi:10.1016/j.jenvp.2022.101904

Bourque, F., & Willox, A. C. (2014). Climate change: The next challenge for public mental health? *International Review of Psychiatry (Abingdon, England), 26*(4), 415–422. doi:10.3109/09540261.2014.925851 PMID:25137107

Brammer, H. (2009). "Climate Refugees": A Rejoinder. *Economic and Political Weekly, 44*(29), 87–87. https://www.jstor.org/stable/40279292

Brennan, S. E., & Munn, Z. (2021). Prisma 2020: A reporting guideline for the next generation of Systematic Reviews. *JBI Evidence Synthesis, 19*(5), 906–908. doi:10.11124/JBIES-21-00112 PMID:33989266

Brocki, L., Dyer, G. C., Gładka, A., & Chung, N. C. (2023). Deep Learning Mental Health Dialogue System. *2023 IEEE International Conference on Big Data and Smart Computing (BigComp)*, 395-398. 10.1109/BigComp57234.2023.00097

Bronen, R. (2013). *Climate-Induced Displacement of Alaska Native Communities, January 30, 2013*. Alaskan Immigration Justice Project. https://www.brookings.edu/wp-content/uploads/2016/06/30-climate-alaska-bronen-paper.pdf

Brown. (2022). Mental Health Impacts of Climate Change: Considerations for Nurse Practitioners. *The Journal for Nurse Practitioners, 18*(4), 359–363. . doi:10.1016/j.nurpra.2021.07.013

Brown, V. J. (2004). BattleScars: Global conflicts and environmental health. *Environmental Health Perspectives, 112*(17), A994–A1003. doi:10.1289/ehp.112-a994 PMID:15579406

Brulle, R. J., Carmichael, J., & Jenkins, J. C. (2012). Shifting public opinion on climate change: An empirical assessment of factors influencing concern over climate change in the US, 2002–2010. *Climatic Change, 114*(2), 653–672. doi:10.1007/s10584-012-0403-y

Buckley, R. (2011). Tourism and environment. *Annual Review of Environment and Resources, 36*(1), 397–416. doi:10.1146/annurev-environ-041210-132637

Buckley, R. (2020). Nature tourism and mental health: Parks, happiness, and causation. *Journal of Sustainable Tourism, 28*(9), 1409–1424. doi:10.1080/09669582.2020.1742725

Buckley, R., & Westaway, D. (2020). Mental health rescue effects of women's outdoor tourism: A role in COVID-19 recovery. *Annals of Tourism Research, 85*, 103041. doi:10.1016/j.annals.2020.103041 PMID:33100433

Burke, M. B., Miguel, E., Satyanath, S., Dykema, J. A., & Lobell, D. B. (2015). Climate change in the Fertile Crescent and implications of the recent Syrian drought. *Proceedings of the National Academy of Sciences of the United States of America, 112*(11), 3241–3246. doi:10.1073/pnas.1421533112 PMID:25733898

Butler, R. W. (2006). *The Tourism Area Life Cycle*. Academic Press.

Butt, T. A., McCarl, B. A., Angerer, J., Dyke, P. T., & Stuth, J. W. (2005). The economic and food security implications of climate change in Mali. *Climatic Change, 68*(3), 355–378. doi:10.1007/s10584-005-6014-0

Calero, C., Bertoa, M. F., & Moraga, M. Á. (2013). A systematic literature review for software sustainability measures. In *2013 2nd international workshop on green and sustainable software (GREENS)*. IEEE. 10.1109/GREENS.2013.6606421

Campbell, J., & Warrick, O. (2014). *Climate change and migration issues in Pacific*. United Nations Economic and Social Commission for Asia and the Pacific. https://www.ilo.org/dyn/migpractice/docs/261/Pacific.pdf

Campbell, B. M., Vermeulen, S. J., Aggarwal, P. K., Corner-Dolloff, C., Girvetz, E., Loboguerrero, A. M., & Wollenberg, E. (2016). Reducing risks to food security from climate change. *Global Food Security, 11*, 34–43. doi:10.1016/j.gfs.2016.06.002

Cantor, D. J. (2018). *Cross-border Displacement, Climate Change and Disasters: Latin America and the Caribbean*. Platform on Disaster Displacement, UNHCR. https://www.unhcr.org/sites/default/files/legacy-pdf/5d4a7b737.pdf

Cazenave, A., & Remy, F. (2011, August 17). Sea level and climate: Measurements and causes of changes. *Wiley Interdisciplinary Reviews: Climate Change, 2*(5), 647–662. doi:10.1002/wcc.139

Central Pollution Control Board (CPCB). (n.d.). *Environmental Protection Act & Rules*. https://cpcb.nic.in/

Ch, Dixit, Joshi, Pant, Thomas, & Beri. (2023). Predicting Employee Mental Health Using Artificial Neural Network. *2023 3rd International Conference on Advance Computing and Innovative Technologies in Engineering (ICACITE)*, 1380-1383. 10.1109/ICACITE57410.2023.10183130

Charanarur, P., Jain, H., Rao, G. S., Samanta, D., Sengar, S. S., & Hewage, C. T. (2023). Machine-Learning-Based Spam Mail Detector. *SN Computer Science*, *4*(6), 858. doi:10.1007/s42979-023-02330-x

Charlson, F., Ali, S., Benmarhnia, T., Pearl, M., Massazza, A., Augustinavicius, J., & Scott, J. G. (2021). Climate change and mental health: A scoping review. *International Journal of Environmental Research and Public Health*, *18*(9), 4486. doi:10.3390/ijerph18094486 PMID:33922573

Chen, T., Song, L., Zhou, H., Li, Y., Wang, H., & Kong, C. (2023). A Study of Mental Health Self-Monitoring Based on the Combination of BERT and Low-Code Platform. *2023 6th World Conference on Computing and Communication Technologies (WCCCT),* 171-175. 10.1109/WCCCT56755.2023.10052392

Chen, C., Chitose, A., Kusadokoro, M., Nie, H., Xu, W., Yang, F., & Yang, S. (2021). Sustainability and challenges in biodiesel production from waste cooking oil: An advanced bibliometric analysis. *Energy Reports*, *7*, 4022–4034. doi:10.1016/j.egyr.2021.06.084

Chen, H., Wang, S., Guo, H., Lin, H., Zhang, Y., Long, Z., & Huang, H. (2019). Study of marine debris around a tourist city in East China: Implication for waste management. *The Science of the Total Environment*, *676*, 278–289. doi:10.1016/j.scitotenv.2019.04.335 PMID:31048159

Cho, E., Um, Y., Yoo, S. K., Lee, H., Kim, H. B., Koh, S., Shin, H. C., & Lee, Y. (2011). An expressed sequence tag analysis for the fast-growing shoots of bambusa edulis murno. *Journal of Plant Biology*, *54*(6), 402–408. doi:10.1007/s12374-011-9179-2

Christensen, J. (2023). *Hot under the collar? Heat can make you angry and even aggressive, research finds.* https://edition.cnn.com/2023/09/05/health/heat-anger-wellness/index.html

Cianconi, P., Betrò, S., & Janiri, L. (2020). The impact of climate change on mental health: A systematic descriptive review. *Frontiers in Psychiatry*, *11*, 74. doi:10.3389/fpsyt.2020.00074 PMID:32210846

Clark-Ginsberg, A., & Chandra, A. (2023). Climate change-related mass migration requires health system resilience. *Environmental Research, Health : ERH*, *1*(4), 045004. doi:10.1088/2752-5309/ace5ca

Clayton, S., & Karazsia, B. (2020). Development and validation of a measure of climate change anxiety. *Journal of Environmental Psychology*, *69*, 101434. doi:10.1016/j.jenvp.2020.101434

Clayton, S., Manning, C. M., Krygsman, K., & Speiser, M. (2021). *Mental health and our changing climate: Impacts, implications, and guidance.* American Psychological Association. doi:10.1037/e507892021-001

Clayton, S., Manning, C., Krygsman, K., & Speiser, M. (2017). *Mental health and our changing climate: impacts, implications, and guidance.* American Psychological Association.

CNBC TV 18. (n.d.). https://tinyurl.com/ye26xhh2

Coffey, Y., Bhullar, N., Durkin, J., Islam, M. S., & Usher, K. (2021, August 1). *Understanding Eco-anxiety: A Systematic Scoping Review of Current Literature and Identified Knowledge Gaps*. The Journal of Climate Change and Health; Elsevier BV. doi:10.1016/j.joclim.2021.100047

Cohen, G. L. (2018). *Climate Change Ethics: Environmental Justice and Global Citizenship*. Rowman & Littlefield.

Cook, C., & Bakker, K. (2012). Water security: Debating an emerging paradigm. *Global Environmental Change*, *22*(1), 94–102. doi:10.1016/j.gloenvcha.2011.10.011

Corvalan, C., Gray, B., Villalobos Prats, E., Sena, A., Hanna, F., & Campbell-Lendrum, D. (2022). Mental health and the global climate crisis. *Epidemiology and Psychiatric Sciences*, *31*, 31. doi:10.1017/S2045796022000361 PMID:36459133

Dahal, R., Samanta, D., Karuppiah, M., & Biswas, J. (2022). Comparative Study of Graph Theory for Network System. International Conference on Innovative Computing and Communications. *Proceedings of ICICC*, *3*, 73–84.

Dalvi, A., Kapur, P., Ashtekar, R., Gorugantu, A., & Siddavatam, I. (2023). Transforming College Counseling: An In-House Solution for Student Mental Wellness. *2023 2nd International Conference for Innovation in Technology (INOCON),* 1-5. 10.1109/INOCON57975.2023.10101364

Daly, P., Dias, A. L., & Patuleia, M. (2020). The impacts of tourism on Cultural Identity on lisbon historic neighbourhoods. *Journal of Ethnic and Cultural Studies*, *8*(1), 1–25. doi:10.29333/ejecs/516

Das & Umamahesh. (2022). Heat Wave Magnitude over India under Changing Climate: Projections from CMIP5 and CMIP6 experiments. *International Journal of Climatology*. doi:10.1002/joc.7246

Dash, L., Pattanayak, B. K., Singh, D., Samanta, D., & Rimal, Y. (2022). Network Lifetime Enhancement by Elimination of Spatially and Temporally Correlated RFID Surveillance Data in WSNs. *Wireless Communications & Mobile Computing*.

Dash, L., Pattanayak, B. K., Singh, D., Samanta, D., & Rimal, Y. (2022). Network Lifetime Enhancement by Elimination of Spatially and Temporally Correlated RFID Surveillance Data in WSNs. *Wireless Communications and Mobile Computing*, *2022*, 1–12. doi:10.1155/2022/2437578

DataHub. (n.d.). *CO2 PPM - Trends in Atmospheric Carbon Dioxide* [Dataset]. https://datahub.io/core/co2-ppm

Davenport, C., & Robertson, C. (2016, May 2). Resettling the First American 'Climate Refugees'. *The New York Times*. https://www.nytimes.com/2016/05/03/us/resettling-the-first-american-climate-refugees.html

Davidson, J. M., & Mill, R. W. (2020). *Mental Health in a Changing Climate: Impacts, Vulnerabilities, and Adaptation*. Springer.

Deepika Manoharan, S. J. (2021, March 17). *Cruise tourism in India: Sailing into troubled waters*. Retrieved from Down To Earth. https://www.downtoearth.org.in/blog/climate-change/cruise-tourism-in-india-sailing-into-troubled-waters-75995

Delello, J. A. (2015). Using Social Media as a Tool for Learning: A Multi-Disciplinary Study. *International Journal on E-Learning*, *14*(2), 163–180.

Desai, Z., & Zhang, Y. (2021). *Climate Change and Women's Health: A Scoping Review*. Academic Press.

DeSalvo, K. B., Hyre, A. D., Ompad, D. C., Menke, A., Tynes, L. L., & Muntner, P. (2007). Symptoms of posttraumatic stress disorder in a new orleans workforce following hurricane Katrina. *Journal of Urban Health*, *84*(2), 142–152. doi:10.1007/s11524-006-9147-1 PMID:17226081

Deshpande, R. S. (2002). Suicide by Farmers in Karnataka: Agrarian Distress and Possible Alleviatory Steps. *Economic and Political Weekly*, *37*(26), 2601–2610. https://www.jstor.org/stable/4412301

Devika, Khademolgorani, Janatyan, & Alavi. (2024). *Smart Waste Management 4.0*. doi:10.1016/j.wasman.2023.08.041

Dierick, D., Hölscher, D., & Schwendenmann, L. (2010). Water use characteristics of a bamboo species (Bambusa Blumeana) in the Philippines. *Agricultural and Forest Meteorology*, *150*(12), 1568–1578. doi:10.1016/j.agrformet.2010.08.006

Doherty, T. J., & Clayton, S. (2016). Climate change and mental health: Causal pathways, policy implications, and research priorities. *EcoHealth*, *13*(4), 654–657.

Duda, A. M., & El-Ashry, M. T. (2000). Addressing the global water and environment crises through integrated approaches to the management of land, water and ecological resources. *Water International*, *25*(1), 115–126. doi:10.1080/02508060008686803

Dunsmore, C. (2022). *Mitigating Mental Health Impacts of Climate-Related Migration: Incorporating psychological resilience building into disaster risk reduction policy frameworks*. Academic Press.

Dwivedi, A. K., Kumar, A., Baredar, P., & Prakash, O. (2019). Bamboo as a complementary crop to address climate change and livelihoods – Insights from India. *Forest Policy and Economics*, *102*(February), 66–74. doi:10.1016/j.forpol.2019.02.007

Easa, N. F. (2019). Social Media Strategies and Students' Satisfaction at Egyptian Universities. *International Journal of Customer Relationship Marketing and Management*, *10*(1), 1–16. doi:10.4018/IJCRMM.2019010101

ECO-ANXIETY on JSTOR. (n.d.). https://doi.org/ doi:10.2307/26798430

Edwards, S., & Harris, M. (Eds.). (2017). *Ethical Challenges in Climate Change Communication: Evolving Perspectives in Science, Policy, and Media*. Springer.

Effects of climate change on future generations. (n.d.). Save the Children. https://www.savethechildren.org/us/what-we-do/emergency-response/climate-change#:~:text=Extreme%20temperatures%20leave%20many%20families,sensitive%20to%20disease%20and%20pollution

Eisenack, K., Moser, S. C., Hoffmann, E., Klein, R. J., Oberlack, C., Pechan, A., Rotter, M., & Termeer, C. J. A. M. (2014). Termeer, C. J: Explaining and overcoming barriers to climate change adaptation. *Nature Climate Change*, *4*(10), 867–872. doi:10.1038/nclimate2350

Eklund, L., Theisen, O. M., Baumann, M., Forø Tollefsen, A., Kuemmerle, T., & Østergaard Nielsen, J. (2022). Societal drought vulnerability and the Syrian climate-conflict nexus are better explained by agriculture than meteorology. *Communications Earth & Environment*, *3*(1), 85. doi:10.1038/s43247-022-00405-w

El-Badawy, T. A., & Hashem, Y. (2015). The Impact of Social Media on the Academic Development of School Students. *International Journal of Business Administration.*

Elezaj, O., Yayilgan, S. Y., Kalemi, E., Wendelberg, L., Abomhara, M., & Ahmed, J. (2019, December). Towards designing a knowledge graph-based framework for investigating and preventing crime on online social networks. In *International Conference on e-Democracy* (pp. 181-195). Springer.

El-Hinnawi, E. (1985). *Environmental Refugees*. United Nations Environmental Programme.

Emamverdian, A., Ding, Y., Ranaei, F., & Ahmad, Z. (2020). Application of bamboo plants in nine aspects. *TheScientificWorldJournal*, *2020*, 1–9. doi:10.1155/2020/7284203 PMID:33061861

Epstein, Y., & Moran, D. S. (2006). Thermal comfort and the heat stress indices. *Industrial Health*, *44*(3), 388–398. doi:10.2486/indhealth.44.388 PMID:16922182

Evans, G. W. (2019). Projected behavioural impacts of global climate change. *Annual Review of Psychology*, *70*(1), 449–474. doi:10.1146/annurev-psych-010418-103023 PMID:29975596

Fareesha. (2023). Real-Time Artificial Mood-tracking and Health-monitoring System (RAMAHS) for people with mental illness and their Caregivers. *2023 International Conference on Recent Trends in Electronics and Communication (ICRTEC)*, 1-6. 10.1109/ICRTEC56977.2023.10111897

Fernandez, A., Black, J., Jones, M., Wilson, L., Salvador-Carulla, L., Astell-Burt, T., & Black, D. (2015). Flooding and mental health: A systematic mapping review. *PLoS One*, *10*(4), e0119929. doi:10.1371/journal.pone.0119929 PMID:25860572

Ferris, E. (2017). Governance and Climate Change-Induced Mobility: International and Regional Frameworks. In D. Manou, A. Baldwin, D. Cubie, A. Mihr, & T. Thorp (Eds.), *Climate Change, Migration and Human Rights: Law and Policy Perspectives* (pp. 49–66). Routledge. doi:10.4324/9781315622217-2

Filho. (2022). Where to Go? Migration and Climate Change Response in West Africa. *Geoforum*, *137*, 83–87. . doi:10.1016/j.geoforum.2022.10.011

Fisher, E. (2016). *Climate Change, Ethics and Human Security*. Cambridge University Press.

Fünfgeld, H. (2015). Facilitating local climate change adaptation through transnational municipal networks. *Current Opinion in Environmental Sustainability*, *12*, 67–73. doi:10.1016/j.cosust.2014.10.011

Füssel, H. M. (2009). An updated assessment of the risks from climate change based on research published since the IPCC Fourth Assessment Report. *Climatic Change, 97*(3-4), 469–482. doi:10.1007/s10584-009-9648-5

Galbraith, J. K., & ProQuest (Firm). (2012). *Inequality and instability: A study of the world economy just before the great crisis* (1st ed.). Oxford University Press. doi:10.1093/acprof:oso bl/9780199855650.001.0001

Gamido & Gamido. (2019). Comparative review of the features of automated software testing tools. *International Journal of Electrical and Computer Engineering, 9*(5).

Garcés-Ordóñez, O., Espinosa Díaz, L. F., Pereira Cardoso, R., & Costa Muniz, M. (2020). The impact of tourism on marine litter pollution on Santa Marta beaches, Colombian Caribbean. *Marine Pollution Bulletin, 160*, 111558. Advance online publication. doi:10.1016/j.marpolbul.2020.111558 PMID:32853837

Garg, M., Kumar, M., & Samanta, D. (2023). *Towards Pattern Recognition with Network Science and Natural Language Processing for Information Retrieval*. Academic Press.

George, G., Howard-Grenville, J., Joshi, A., & Tihanyi, L. (2016). Understanding and tackling societal grand challenges through management research. *Academy of Management Journal, 59*(6), 1880–1895. doi:10.5465/amj.2016.4007

George, G., Merrill, R. K., & Schillebeeckx, S. J. (2021). Digital sustainability and entrepreneurship: How digital innovations are helping tackle climate change and sustainable development. *Entrepreneurship Theory and Practice, 45*(5), 999–1027. doi:10.1177/1042258719899425

Ghose, A., Bharadwaj, S., & Aamir Ali, S. M. (2023). Politics of Slaughter: A Critical Review of Constitutional Failure in Myanmar and India During Rohingya Crisis. *Novum Jus, 17*(2), 251–277. doi:10.14718/NovumJus.2023.17.2.10

Giovanis, E., & Abdeldjalil, A A. (2023). The Impact of Climate Change-Induced Migration on Well-being In Joint Policy Study. *Climate-Induced Internal Displacement in MENA Countries.*

Gladwin, C. H., Thomson, A. M., Peterson, J. S., & Anderson, A. S. (2001). Addressing food security in Africa via multiple livelihood strategies of women farmers. *Food Policy, 26*(2), 177–207. doi:10.1016/S0306-9192(00)00045-2

Global Compact for Migration. (2018). *Global Compact for Safe, Orderly and Regular Migration.* https://refugeesmigrants.un.org/sites/default/files/180713_agreed_outcome_global_compact_for_migration.pdf

Goodwin, H. (2017). *The challenge of overtourism.* Responsible Tourism Partnership Working Paper 4. October.

Goodwin, J., & Zaman, U. (2023). Mental health stigma and UN Sustainable Development Goals. *Frontiers in Psychiatry, 14*, 1190406. doi:10.3389/fpsyt.2023.1190406 PMID:37091707

Gordon, B. (1986). The souvenir: Messenger of the extraordinary. *Journal of Popular Culture, 20*(3), 135–146. doi:10.1111/j.0022-3840.1986.2003_135.x

Gössling, S., & Michael Hall, C. (2019). Sharing versus collaborative economy: How to align ICT developments and the SDGs in tourism? *Journal of Sustainable Tourism, 27*(1), 74–96. Advance online publication. doi:10.1080/09669582.2018.1560455

Goudet, J.-M., Arif, F. B., Owais, H., Ahmed, H. U., & Ridde, V. (2023). *Intersecting Vulnerabilities: Unraveling the Complex Relationship between Women.* Climate Change, and Mental Health in Bangladeshi Communities.

Government of India. (n.d.). https://tinyurl.com/m8cxvhtn

Goyal, F., Bhambri, P., & Bansal, P. (2012). Accuracy of Phylogenetic Trees using Distance based Methods. *International Conference on Sports Biomechanics, Emerging Technologies and Quality Assurance in Technical Education*, 400-403.

Grasser, L. R. (2022). Addressing Mental Health Concerns in Refugees and Displaced Populations: Is Enough Being Done? *Risk Management and Healthcare Policy, 15*, 909–922. . doi:10.2147/RMHP.S270233

Green, B., & Johnson, M. (2020). *The Ethics of Climate Change: An Introduction.* Routledge.

Gregory, P. J., Ingram, J. S., & Brklacich, M. (2005). Climate change and food security. *Philosophical Transactions of the Royal Society of London. Series B, Biological Sciences, 360*(1463), 2139–2148. doi:10.1098/rstb.2005.1745 PMID:16433099

Grose, A. (2020, June 9). *A Guide to Eco-Anxiety.* Watkins Publishing.

Guiney, R. (2012). Farming suicides during the Victorian drought: 2001–2007. *The Australian Journal of Rural Health, 20*(1), 11–15. doi:10.1111/j.1440-1584.2011.01244.x PMID:22250871

Gunasiri, H., Wang, Y., Watkins, E., Capetola, T., Henderson-Wilson, C., & Patrick, R. (2022). Hope, Coping and Eco-Anxiety: Young People's Mental Health in a Climate-Impacted Australia. *International Journal of Environmental Research and Public Health, 19*(9), 5528. doi:10.3390/ijerph19095528 PMID:35564923

Gupta, S., Nischal, P., & Bhmabri, P. (2007). Data Encryption Standard (DES) Algorithm with Diffie-Hellman Key Exchange. *National Conference on Emerging Trends in Communication & IT,* 135-143.

Gupta, V., Joshi, V., Jain, A., & Garg, I. (2023). Chatbot for Mental health support using NLP. *2023 4th International Conference for Emerging Technology (INCET),* 1-6. 10.1109/INCET57972.2023.10170573

Gu, W., Lou, G., Tan, W., & Yuan, X. (2017, November). A Nonlinear State Estimator-Based Decentralized Secondary Voltage Control Scheme for Autonomous Microgrids. *IEEE Transactions on Power Systems, 32*(6), 4794–4804. doi:10.1109/TPWRS.2017.2676181

Haites, E. (2009). Linking emissions trading schemes for international aviation and shipping emissions. *Climate Policy*, *9*(4), 415–430. doi:10.3763/cpol.2009.0620

Hanigan, I. C., Butler, C. D., Kokic, P. N., & Hutchinson, M. F. (2012). Suicide and drought in new south wales, australia, 1970–2007. *Proceedings of the National Academy of Sciences of the United States of America*, *109*(35), 13950–13955. doi:10.1073/pnas.1112965109 PMID:22891347

Hannart, A., & Naveau, P. (2018, July). Probabilities of Causation of Climate Changes. *Journal of Climate*, *31*(14), 5507–5524. doi:10.1175/JCLI-D-17-0304.1

Hansen, J., & Sato, M. (2012). Assessing "Dangerous Climate Change": Required Reduction of Carbon Emissions to Protect Young People, Future Generations and Nature. *PLoS One*, *7*(12), e51068. PMID:23251423

Hayes, K., Blashki, G., Wiseman, J., Burke, S., & Reifels, L. (2018). Climate change and mental health: Risks, impacts and priority actions. *International Journal of Mental Health Systems*, *12*(1), 28. doi:10.1186/s13033-018-0210-6 PMID:29881451

Hepp, J., Klein, S. A., Horsten, L. K., Urbild, J., & Lane, S. P. (2022). The Climate Change Distress and Impairment Scale: Introduction of the Measure and First Findings on pro-. *Environment and Behavior*.

Hickman, C. (2020, October 1). We need to (find a way to) talk about . . . Eco-anxiety. *Journal of Social Work Practice*, *34*(4), 411–424. doi:10.1080/02650533.2020.1844166

Hinds, J., & Sparks, P. (2017). Mental health impacts of floods. *Current Psychiatry Reports*, *19*(11), 1–10.

Hipp, J. R., Bates, C., Lichman, M., & Smyth, P. (2019). Using social media to measure temporal ambient population: Does it help explain local crime rates? *Justice Quarterly*, *36*(4), 718–748. doi:10.1080/07418825.2018.1445276

Hogg, T. L., Stanley, S. K., O'Brien, L. V., Wilson, M. S., & Watsford, C. R. (2021). The Hogg Eco-Anxiety Scale: Development and validation of a multidimensional scale. *Global Environmental Change*, *71*(102391), 1–10. doi:10.1016/j.gloenvcha.2021.102391

Holmes, A., & Kellogg, M. (2006). *Automating functional tests using selenium. AGILE 2006 (AGILE'06).*

Howatt, B. (2020, February 12). How does climate-driven eco-anxiety affect our performance at work? *The Globe and Mail*. https://www.theglobeandmail.com/business/careers/workplace-award/article-how-does-climate-driven-eco-anxiety-affect-our-performance-at-work/

Hrabok, M., Delorme, A., & Agyapong, V. I. (2020). Threats to mental health and well-being associated with climate change. *Journal of Anxiety Disorders*, *76*, 102295. doi:10.1016/j.janxdis.2020.102295 PMID:32896782

Hsiang, S. M., & Burke, M. (2014). Climate, conflict, and social stability: What does the evidence say? *Climatic Change*, *123*(1), 39–55. doi:10.1007/s10584-013-0868-3

Hu, C.-P., Hu, Y., & Yan, W.-W. (2014, October). An empirical study of factors influencing user perception of university digital libraries in China. *Library & Information Science Research*, *36*(3-4), 225–233. doi:10.1016/j.lisr.2013.10.008

Human Rights Council. (2009). *Tenth Session. Resolution 10/4. Human rights and climate change.* https://ap.ohchr.org/documents/E/HRC/resolutions/A_HRC_RES_10_4.pdf

Hwong, A. R., Wang, M., Khan, H., Chagwedera, D. N., Grzenda, A., Doty, B., Benton, T., Alpert, J., Clarke, D., & Compton, W. M. (2022). Compton, W. M.: Climate change and mental health research methods, gaps, and priorities: A scoping review. *The Lancet. Planetary Health*, *6*(3), e281–e291. doi:10.1016/S2542-5196(22)00012-2 PMID:35278392

Identifying Types of Eco-Anxiety, Eco-Guilt, Eco-Grief, and Eco-Coping in a Climate-Sensitive Population: A Qualitative Study. (n.d.). International Journal of Environmental Research and Public Health; Multidisciplinary Digital Publishing Institute. doi:10.3390/ijerph19042461

Ilahl, R. a. (2019). Digital libraries as learning resources. *Journal of Physics: Conference Series*, *1402*, 1–6.

India Government. (n.d.). *Environment and Forest*. https://tinyurl.com/2s5ch437

Indian Council of Forestry Research and Education. (n.d.). https://icfre.gov.in/

IPCC. (2018). *Global Warming of 1.5°C. An IPCC Special Report on the impacts of global warming of 1.5°C above pre-industrial levels and related global greenhouse gas emission pathways, in the context of strengthening the global response to the threat of climate change, sustainable development, and efforts to eradicate poverty*. Retrieved from https://www.ipcc.ch/sr15/

Ippolito, M., & Cicatiello, L. (2019). Political instability, economic inequality and social conflict: The case in italy. *Panoeconomicus*, *66*(3), 365–383. doi:10.2298/PAN190336SI

Israr, H. A., Yap, C. T. M., Wong, K. J., & Yahya, M. Y. (2016). Compressive properties of Hawaiian Gold Timber Bamboo under different conditions. *Journal of Advanced Research in Applied Mechanics*, *25*(1), 10–18.

Jage, S., Chaudhari, S., Jatte, M., Mhatre, A., & Mane, V. (2023). Predicting Mental Health Illness using Machine Learning. *2023 3rd Asian Conference on Innovation in Technology (ASIANCON)*, 1-5. 10.1109/ASIANCON58793.2023.10270445

Jamieson, D. (2014). *Reason in a Dark Time: Why the Struggle to Stop Climate Change Failed—And What It Means for Our Future*. Oxford University Press. doi:10.1093/acprof:oso/9780199337668.001.0001

Jessee, N. (2020). Community resettlement in louisiana: Learning from histories of horror and hope. In S. Laska (Ed.), *Louisiana's Response to Extreme Weather* (pp. 147–184). Springer International Publishing. doi:10.1007/978-3-030-27205-0_6

Johann, T. (2011). Sustainable development, sustainable software, and sustainable software engineering: an integrated approach. In *2011 International Symposium on Humanities, Science and Engineering Research.* IEEE. 10.1109/SHUSER.2011.6008495

Jones, A. (2022). The Health Impacts of Climate Change: Why Climate Action Is Essential to Protect Health. *Orthopaedics and Trauma, 36*(5), 248–255. doi:10.1016/j.mporth.2022.07.001

Jones, J. W., Hansen, J. W., Royce, F. S., & Messina, C. D. (2000). Potential benefits of climate forecasting to agriculture. *Agriculture, Ecosystems & Environment, 82*(1-3), 169–184. doi:10.1016/S0167-8809(00)00225-5

Jusoh, S., Kadir, R. A., & Al Fawareh, H. (2023). mentBot: A Repository for Mental Health Disorders. *2023 International Conference on Consumer Electronics - Taiwan (ICCE-Taiwan),* 153-154. 10.1109/ICCE-Taiwan58799.2023.10227068

Jyoti Nath, A., & Das, A. K. (2012). Ecological implications of village bamboo as global climate change mitigation strategy. *International Journal of Climate Change Strategies and Management, 4*(2), 201–215. doi:10.1108/17568691211223169

Kari, H. K. (2019). Promoting library services the social media era exploring use objectives challenges and percieved influence. *Philosophy of Education.*

Karita, L., Mourão, B. C., & Machado, I. (2019). Software industry awareness on green and sustainable software engineering: a state-of-the-practice survey. *Proceedings of the XXXIII Brazilian Symposium on Software Engineering*. 10.1145/3350768.3350770

Kartha, J. R., Dahal, R., Samanta, D., Poonia, R. C., Bhattacharya, A., & Dutta, S. (2022). A Shortest Path Problem for Drug Delivery Using Domination and Eccentricity. In Human-Centric Smart Computing [Springer Nature Singapore Singapore.]. *Proceedings of ICHCSC, 2022*, 207–215.

Karthi, M., Alsager, M., Metha, R., Fatima, N. N., & Al-Gindy, A. (2023). Digital Solution: Breaking the Barriers to Address Stigma of Mental Health. *IEEE EUROCON 2023 - 20th International Conference on Smart Technologies*, 24-29. 10.1109/EUROCON56442.2023.10198879

Kaur, A., Bhambri, P., & Gupta, O. P. (2012). A Novel Technique for Robust Image Segmentation. *International Journal of Advances in Engineering and Technology, 3*(4), 110–114.

Kaur, P., & Bhambri, P. (2013). Review of Zero Watermarking Algorithms for Text Documents. *PCTE Journal of Computer Sciences, 10*(1), 77–82.

Kaushik, A. (2022, October 28). Eco anxiety. *British Dental Journal, 233*(8), 588–588. doi:10.1038/s41415-022-5164-z PMID:36307681

Kelman, I., Orlowska, J., Upadhyay, H., Stojanov, R., Webersik, C., Simonelli, A. C., Procházka, D., & Němec, D. (2019). Does climate change influence people's migration decisions in Maldives? *Climatic Change, 153*(1–2), 285–299. doi:10.1007/s10584-019-02376-y

Kerz, E., Wiechmann, D., & Riedel, F. B. (2017). Implicit learning in the crowd. *Studies in Second Language Acquisition, 39*(4), 711–734. doi:10.1017/S027226311700002X

Khan, R. (2021, October- December). Importance of digital library in education. *International Journal of Research in Library Science*, *7*(4), 102–117. doi:10.26761/IJRLS.7.4.2021.1467

Kilgo, D. K., Harlow, S., García-Perdomo, V., & Salaverría, R. (2018). A new sensation? An international exploration of sensationalism and social media recommendations in online news publications. *Journalism*, *19*(11), 1497–1516. doi:10.1177/1464884916683549

Kjellstrom, T., Kovats, R. S., Lloyd, S. J., Holt, T., & Tol, R. S. J. (2009). The Direct Impact of Climate Change on Regional Labor Productivity. *Archives of Environmental & Occupational Health*, *64*(4), 217–227. doi:10.1080/19338240903352776 PMID:20007118

Koens, K., Postma, A., & Papp, B. (2018). Is overtourism overused? Understanding the impact of tourism in a city context. *Sustainability (Basel)*, *10*(12), 4384. doi:10.3390/su10124384

Koopman, C., Classen, C., Cardeña, E., & Spiegel, D. (1995, January). When disaster strikes, acute stress disorder may follow. *Journal of Traumatic Stress*, *8*(1), 29–46. doi:10.1002/jts.2490080103 PMID:7712057

Kostopoulou, E., & Giannakopoulos, C. (2023, February 21). Projected Changes in Extreme Wet and Dry Conditions in Greece. *Climate (Basel)*, *11*(3), 49. doi:10.3390/cli11030049

Kumar, K., Chaudhary, S., Anandaram, H., Kumar, R., Gupta, A., & Joshi, K. (2023). Industry 4.0 and Health Care System with special reference to Mental Health. *2023 1st International Conference on Intelligent Computing and Research Trends (ICRT)*, 1-5. 10.1109/ICRT57042.2023.10146640

Kumar, K. K., Prasanna, V. L., & Roy, P. S. (2022). Floods in South India: An overview of causes and consequences. In *Climate Change Impacts and Challenges in India* (pp. 151–165). Springer.

Lal, R. (2004). Soil carbon sequestration impacts on global climate change and food security. *Science*, *304*(5677), 1623–1627. doi:10.1126/science.1097396 PMID:15192216

Lal, R. (2007). Carbon sequestration. *Philosophical Transactions of the Royal Society of London. Series B, Biological Sciences*, *363*(1492), 815–830. doi:10.1098/rstb.2007.2185 PMID:17761468

Lawrance, E. L., Thompson, R., Newberry Le Vay, J., Page, L., & Jennings, N. (2022). The impact of climate change on mental health and emotional wellbeing: A narrative review of current evidence, and its implications. *International Review of Psychiatry (Abingdon, England)*, *34*(5), 443–498. doi:10.1080/09540261.2022.2128725 PMID:36165756

Lázár, A. N., Nicholls, R. J., Hall, J. W., Barbour, E. J., & Haque, A. (2020). Contrasting development trajectories for coastal Bangladesh to the end of century. *Regional Environmental Change*, *20*(3), 93. doi:10.1007/s10113-020-01681-y

Lebel, L., Paquin, V., Kenny, T. A., Fletcher, C., Nadeau, L., Chachamovich, E., & Lemire, M. (2022). Climate change and Indigenous mental health in the Circumpolar North: A systematic review to inform clinical practice. *Transcultural Psychiatry*, *59*(3), 312–336. doi:10.1177/13634615211066698 PMID:34989262

Lê, F., Tracy, M., Norris, F. H., & Galea, S. (2013). Displacement, county social cohesion, and depression after a large-scale traumatic event. *Social Psychiatry and Psychiatric Epidemiology*, *48*(11), 1729–1741. doi:10.1007/s00127-013-0698-7 PMID:23644724

Legg, T. J., & Raypole, C. (2020). *Climate Change Taking a Toll on Your Mental Health?* https://www.healthline.com/health/eco-anxiety

Leiserowitz, A., Maibach, E., Rosenthal, S., Kotcher, J., Bergquist, P., Ballew, M., Goldberg, M., & Gustafson, A. (2019). *Climate Change in the American Mind: April 2019*. Yale University and George Mason University. Retrieved from https://climatecommunication.yale.edu/publications/climate-change-in-the-american-mind-april-2019/

Lelean, J. (2021, November 8). The Maldives is threatened by rising seas – but coastal development is causing even more pressing environmental issues. *Connecting Research*. https://research.reading.ac.uk/research-blog/the-maldives-is-threatened-by-rising-seas-but-coastal-development-is-causing-even-more-pressing-environmental-issues/

Lenzen, M., Sun, Y. Y., Faturay, F., Ting, Y. P., Geschke, A., & Malik, A. (2018). The carbon footprint of global tourism. *Nature Climate Change*, *8*(6), 522–528. doi:10.1038/s41558-018-0141-x

Lessard, G., & Chouinard, A. (1980). Bamboo research in Asia. In *Proceedings of a workshop held in Singapore,* 28-30 May 1980. International Development Research Centre.

Li, Y. (2023). Application of Machine Learning to Predict Mental Health Disorders and Interpret Feature Importance. *2023 3rd International Symposium on Computer Technology and Information Science (ISCTIS),* 257-261. 10.1109/ISCTIS58954.2023.10213032

Liao. (2023). Association between Internal Migration Experience and Depressive Symptoms: Analysis of PSID Data. *BMC Public Health, 23*(1). . doi:10.1186/s12889-023-16073-0

Li, C., Lawrance, E. L., Morgan, G., Brown, R., Greaves, N., Krzanowski, J., Samuel, S., Guinto, R. R., & Belkin, G. (2022). Belkin, G.: The role of mental health professionals in the Climate Crisis: An urgent call to action. *International Review of Psychiatry (Abingdon, England)*, *34*(5), 563–570. doi:10.1080/09540261.2022.2097005 PMID:36165755

Liubov Zharova. (2021). *4 Sustainable Development Goals as a Roadmap for Territorial Development," in Economic Inequality – Trends, Traps and Trade-offs*. River Publishers.

Liu, S., Yang, L., Zhang, C., Xiang, Y. T., Liu, Z., Hu, S., & Zhang, B. (2020). Online mental health services in China during the COVID-19 outbreak. *The Lancet. Psychiatry*, *7*(4), e17–e18. doi:10.1016/S2215-0366(20)30077-8 PMID:32085841

Liu, S., & Young, S. D. (2018). A survey of social media data analysis for physical activity surveillance. *Journal of Forensic and Legal Medicine*, *57*, 33–36. doi:10.1016/j.jflm.2016.10.019 PMID:29801949

Llorente-Marrón, M., Díaz-Fernández, M., Méndez-Rodríguez, P., & González Arias, R. (2020). Social vulnerability, gender and disasters. The case of Haiti in 2010. *Sustainability (Basel), 12*(9), 3574. doi:10.3390/su12093574

Loehr, J., Becken, S., Nalau, J., & Mackey, B. (2022). Exploring the multiple benefits of ecosystem-based adaptation in tourism for climate risks and destination well-being. *Journal of Hospitality & Tourism Research (Washington, D.C.), 46*(3), 518–543. doi:10.1177/1096348020944438

Lohia, A., Ranjith, A., Sachdeva, A., & Bhat, A. (2023). An Ensemble Technique to Analyse Mental Health of Workforce in the Corporate Sector. *2023 7th International Conference on Intelligent Computing and Control Systems (ICICCS),* 1245-1251. 10.1109/ICICCS56967.2023.10142470

Lokotola, C. L. (2023). Towards a climate-resilient primary health care service. *South African Family Practice, 65*(1). . doi:10.4102/safp.v65i1.5749

Lubega, D. P. (2019). *Leveraging Social Media in Higher Education: A Case of universities in Uganda.* European Journal of Open, Distance and e-Learning.

Lu, C.-W., Huang, J.-C., Chen, C., Shu, M.-H., Hsu, C.-W., & Tapas Bapu, B. R. (2021). An energy-efficient smart city for Sustainable Green Tourism Industry. *Sustainable Energy Technologies and Assessments, 47,* 101494. doi:10.1016/j.seta.2021.101494

Luo, X., Wang, Q., Zheng, B., Lin, L., Chen, B., Zheng, Y., & Xiao, J. (2017). Hydration properties and binding capacities of dietary fibers from bamboo shoot shell and its hypolipidemic effects in mice. *Food and Chemical Toxicology, 109*(2), 1003–1009. doi:10.1016/j.fct.2017.02.029 PMID:28237776

Lutz, J., & Yayboke, E. (2021, November 22). *Haitian Migration: Food Insecurity, and a Better Way Forward.* Centre for Strategic & International Studies. https://www.csis.org/analysis/haitian-migration-food-insecurity-fragility-and-better-way-forward

M, S. N., & Khan, F. R. (2019). Examining the impact of social media on the academic performances of Saudi students Prince Sattam Bin Abdul University. *Humanities & Social Sciences Reviews.*

Mahmoud, S. S., & Ahmad, I. (2013). A green model for sustainable software engineering. *International Journal of Software Engineering and Its Applications, 7*(4), 55–74.

Maldonado, J., Wang, I. F. C., Eningowuk, F., Iaukea, L., Lascurain, A., Lazrus, H., Naquin, C. A., Naquin, J., Nogueras-Vidal, K. M., Peterson, K., Rivera-Collazo, I., Souza, M. K., Stege, M., & Thomas, B. (2021). Addressing the challenges of climate-driven community-led resettlement and site expansion: Knowledge sharing, storytelling, healing, and collaborative coalition building. *Journal of Environmental Studies and Sciences, 11*(3), 294–304. doi:10.1007/s13412-021-00695-0 PMID:34099966

Malleson, N., & Andresen, M. A. (2015). The impact of using social media data in crime rate calculations: Shifting hot spots and changing spatial patterns. *Cartography and Geographic Information Science, 42*(2), 112–121. doi:10.1080/15230406.2014.905756

Mange, D., Vaidyanathan, M., Lanka, S. P., & Ahuja, S. (2023). Application for performing self care activities using flutter. *2023 2nd International Conference on Vision Towards Emerging Trends in Communication and Networking Technologies (ViTECoN).*

Marandi, A. K., & Shah, H. (2023). Treatment of Mental Health Disorders Advanced By Artificial Intelligence. *2023 3rd International Conference on Advance Computing and Innovative Technologies in Engineering (ICACITE),* 2569-2573. 10.1109/ICACITE57410.2023.10182496

Marazziti. (2021). Climate Change, Environment Pollution, COVID-19 Pandemic and Mental Health. *Science of The Total Environment, 773,* 145182. doi:10.1016/j.scitotenv.2021.145182

Mathers-Jones, J., & Todd, J. (2023). Ecological anxiety and pro-environmental behaviour: The role of attention. *Journal of Anxiety Disorders, 98,* 102745. doi:10.1016/j.janxdis.2023.102745 PMID:37480627

McAdam, J. (2012). *Climate Change, Forced Migration, and International Law.* Oxford University Press. doi:10.1093/acprof:oso/9780199587087.001.0001

McIver. (2016). Health Impacts of Climate Change in Pacific Island Countries: A Regional Assessment of Vulnerabilities and Adaptation Priorities. *Environmental Health Perspectives, 124*(11), 1707–1714. doi:10.1289/ehp.1509756

McLeman, R. A., & Hunter, L. M. (2010). Migration in the context of vulnerability and adaptation to climate change: Insights from analogues. *Wiley Interdisciplinary Reviews: Climate Change, 1*(3), 450–461. doi:10.1002/wcc.51 PMID:22022342

McMichael, A. J., & Lindgren, E. (2011). Climate change: Present and future risks to health, and necessary responses. *Journal of Internal Medicine, 270*(5), 401–413. doi:10.1111/j.1365-2796.2011.02415.x PMID:21682780

McMichael, C. (2023). Climatic and environmental change, migration, and health. *Annual Review of Public Health, 44*(1), 171–191. doi:10.1146/annurev-publhealth-071421-045148 PMID:36542773

McMichael, C., Barnett, J., & McMichael, A. J. (2012). An ill wind? Climate change, migration, and health. *Environmental Health Perspectives, 120*(5), 646–654. doi:10.1289/ehp.1104375 PMID:22266739

McMillen, C., North, C., Mosley, M., & Smith, E. (2002). Untangling the psychiatric comorbidity of posttraumatic stress disorder in a sample of flood survivors. *Comprehensive Psychiatry, 43*(6), 478–485. doi:10.1053/comp.2002.34632 PMID:12439837

Medani, P. (2021a). *1 Trends, Traps and Trade-offs of Economic Inequality—A Growing Pattern and Concern, in Economic Inequality – Trends, Traps and Trade-offs.* River Publishers.

Medani, P. (2021b). *12 Conclusion—The Unpredictable Direction of Economic Inequality, in Economic Inequality – Trends, Traps and Trade-offs.* River Publishers.

Medani, P. (2021c). *Authors Biographies, in Economic Inequality – Trends, Traps and Trade-offs.* River Publishers.

Medani, P. (2021d). *Economic Inequality – Trends, Traps and Trade-offs, in Economic Inequality – Trends, Traps and Trade-offs*. River Publishers.

Medani, P. (2021e). *Word Phrase Definitions and Abbreviation for Economic Inequality – Trends, Traps and Trade-offs, in Economic Inequality – Trends, Traps and Trade-offs*. River Publishers.

Merrill, Schillebeeckx, & Blakstad. (2019). *Sustainable digital finance in Asia: Creating environmental impact through bank transformation*. SDFA, DBS, UN Environment.

Methmann, C., & Oels, A. (2015). From 'fearing' to 'empowering' climate refugees: Governing climate-induced migration in the name of resilience. *Security Dialogue*, *46*(1), 51–68. doi:10.1177/0967010614552548

Miguel, C. S., Morales, K., & Ynalvez, M. A. (2020). Online Victimization, Social Media Utilization, and Cyber Crime Prevention Measures. *Asia-Pacific Social Science Review, 20*(4).

Mikalef, P., Bourab, M., Lekakos, G., & Krogstie, J. (2020). The role of information governance in big data analytics driven innovation. *Information & Management*, *57*(7), 103361. doi:10.1016/j.im.2020.103361

Milivojevic, S., & McGovern, A. (2014). The death of Jill Meagher: Crime and punishment on social media. *International Journal for Crime, Justice and Social Democracy, 3*(3), 22-39.

Miller. (2020). *Climate change solutions: The role of technology*. House of Commons Library.

Miller, G. (2017). Climate Change and Mental Health: A Scoping Review. *International Journal of Mental Health Systems*, *11*, 43.

Miller, V. J., Maziarz, L., Wagner, J., Bell, J., & Burek, M. (2023). Nursing assistant turnover in nursing homes: A scoping review of the literature. *Geriatric Nursing*, *51*, 360–368. doi:10.1016/j.gerinurse.2023.03.027 PMID:37104907

Milman, O. (2016, December, 19). Alaska indigenous people see culture slipping away as sea ice vanishes. *The Guardian*. https://www.theguardian.com/environment/2016/dec/19/alaska-sea-ice-vanishing-climate-change-indigenous-people

Mishra, C. (2019). *Impact of Social Media on Academic Business School Libraries in India: An Empirical Study*. Library Philosophy and Practice.

Mishra, N., & Nagaraj, K. (2016). Suicides of farmers in Maharashtra, India: A qualitative exploration of their causes. *Journal of Injury & Violence Research*, *8*(1).

Mishra, S. (2014). Agrarian distress and farmers' suicides in India: Is there any way out? *Indian Journal of Agricultural Economics*, *69*(3), 375–388.

Modi Yojanaye, P. M. (n.d.). *Jal Jeevan Mission (Rural) Scheme*. https://tinyurl.com/axmrtx2s

Mohammadian, H. D., Wittberg, V., Castro, M., & Bolandian, G. (2020). *The 5th Wave and i-Sustainability Plus Theories as Solutions for SocioEdu Consequences of Covid-19. In 2020 IEEE Learning With MOOCS*. LWMOOCS., doi:10.1109/LWMOOCS50143.2020.9234360

Mora, C., Rollins, R. L., Taladay, K., Kantar, M. B., Chock, M. K., Shimada, M., & Franklin, E. C. (2018). Bitcoin emissions alone could push global warming above 2C. *Nature Climate Change, 8*(11), 931–933. doi:10.1038/s41558-018-0321-8

Mostafanezhad, M., & Evrard, O. (2018). Geopolitical ecologies of tourism and the transboundary haze disaster in Thailand, Laos and myanmar. Community, Environment and Disaster Risk Management, 19. doi:10.1108/S2040-726220180000019003

Mozaffar, H., & Panteli, N. (2021). The online community knowledge flows: Distance and direction. *European Journal of Information Systems*, 1–14.

Müller, K., & Schwarz, C. (2021). Fanning the flames of hate: Social media and hate crime. *Journal of the European Economic Association, 19*(4), 2131–2167. doi:10.1093/jeea/jvaa045

NAB. (n.d.). https://tinyurl.com/4sb36wy9

Nadeem, M., Rashid, J., Moon, H., & Dosset, A. (2023). Machine Learning for Mental Health: A Systematic Study of Seven Approaches for Detecting Mental Disorders. *2023 International Technical Conference on Circuits/Systems, Computers, and Communications (ITC-CSCC),* 1-6. 10.1109/ITC-CSCC58803.2023.10212609

Nagarajaiah, K., Krishnappa, M. H., & Asha, K. R. (2023). Machine Learning based Detection of Post Traumatic Stress Disorder of Mental Health. *2023 4th International Conference on Electronics and Sustainable Communication Systems (ICESC),* 1456-1460. 10.1109/ICESC57686.2023.10193036

Namaziandost, E., & Nasri, M. (2019). The Impact of Social Media on EFL Learners' Speaking Skill A Survey Study Involving EFL Teachers and Students. *Journal of Applied Linguistics and Language Research.*

National Biodiversity Authority (NBA). (n.d.). *Know the Biological Diversity Act 2002 and the rules 2004, Information on Biodiversity.* https://tinyurl.com/5n7tb2ve

National Crime Records Bureau. (2020). *Accidental deaths & suicides in India (ADSI) report-2019 & ADSI report-2020 (provisional data).* Ministry of Home Affairs.

National Disaster Management Authority (NDMA). (n.d.). *Natural.* https://ndma.gov.in/

Nau, M. D. (2016). *Whose financial crisis? how the great recession reshaped economic instability and inequality in the U.S.* Academic Press.

Neethu & Ramesh. (2023). *Projected Changes in Heat Wave Characteristics Over India*. Springer.

Nguyen, H. Q. (2001). *Testing applications on the Web: Test planning for Internet-based systems. John Wiley & Sons*. Web Application Testing.

Nicholas. (2020). Mental Health Impacts of Climate Change: Perspectives for the ED Clinician. *Journal of Emergency Nursing, 46*(5), 590–599. doi:10.1016/j.jen.2020.05.014

Nickerson. (2019). The Association between Visa Insecurity and Mental Health, Disability and Social Engagement in Refugees Living in Australia. *European Journal of Psychotraumatology, 10*(1). doi:10.1080/20008198.2019.1688129

Nisbet, M. C. (2018). Communicating Climate Change: Why Frames Matter for Public Engagement. *Environment, 60*(4), 12–23.

Nitschke, M., Tucker, G. R., Hansen, A. L., Williams, S., Zhang, Y., & Bi, P. (2011). Impact of two recent extreme heat episodes on morbidity and mortality in Adelaide, South Australia: A case-series analysis. *Environmental Health, 10*(1), 42. doi:10.1186/1476-069X-10-42 PMID:21592410

Norgaard, K. M. (2011). *Living in denial: Climate change, emotions, and the denial of grief.* Oxford University Press. doi:10.7551/mitpress/9780262015448.001.0001

Noriega, M. (2020). The application of artificial intelligence in police interrogations: An analysis addressing the proposed effect AI has on racial and gender bias, cooperation, and false confessions. *Futures, 117*. doi:10.1016/j.futures.2019.102510

Nouman, M., Sara, H., Khoo, S. Y., Mahmud, M. P., & Kouzani, A. Z. (2023). Mental Health Prediction through Text Chat Conversations. *2023 International Joint Conference on Neural Networks (IJCNN),* 1-6. 10.1109/IJCNN54540.2023.10191849

Novera, C. N., & Connolly, R., & P. W. (2006). Past, present and future impact of social media on health workers' mental health: A text mining approach. *Journal of Modelling in Management.*

Ojo, A. O., & Fauzi, M. A. (2020). Environmental awareness and leadership commitment as determinants of IT professionals engagement in Green IT practices for environmental performance. *Sustainable Production and Consumption, 24*, 24. doi:10.1016/j.spc.2020.07.017

Osintseva, A. A., & Gvozdetskaya, S. S. (2023). ICT and Mental Health of Professionals Working from Home. *2023 IEEE 24th International Conference of Young Professionals in Electron Devices and Materials (EDM),* 1980-1984. 10.1109/EDM58354.2023.10225054

Pachauri, R. K., & Reisinger, A. (Eds.). (2007). *Climate Change 2007: Synthesis Report. Contribution of Working Groups I, II, and III to the Fourth Assessment Report of the Intergovernmental Panel on Climate Change*. IPCC.

Padhy, S., Sarkar, S., Panigrahi, M., & Paul, S. (2015). Mental health effects of climate change. *Indian Journal of Occupational and Environmental Medicine, 19*(1), 3. doi:10.4103/0019-5278.156997 PMID:26023264

Paika, V., & Bhambri, P. (2013). Edge Detection Technique based on Fuzzy Logic. *International Journal of Mathematics and Computer Research, 1*(3), 83–87.

Palinkas, L. A., O'Donnell, M. L., Lau, W., & Wong, M. (2020). Strategies for delivering mental health services in response to global climate change: A narrative review. *International Journal of Environmental Research and Public Health, 17*(22), 8562. doi:10.3390/ijerph17228562 PMID:33218141

Pandi, P. P. (2020). *Impact and Usage of Social Media Among the Post graduate students of Alagappa university Karaikudi.* Research gate.

Pandve, H. T. (2009). India's national action plan on climate change. *Indian Journal of Occupational and Environmental Medicine*, *13*(1), 17. doi:10.4103/0019-5278.50718 PMID:20165607

Pang, S. F., McKercher, B., & Prideaux, B. (2013). Climate change and tourism: An overview. *Asia Pacific Journal of Tourism Research*, *18*(1–2), 4–20. doi:10.1080/10941665.2012.688509

Papakyriakopoulos, O., & Zuckerman, E. (2021, May). The Media During the Rise of Trump: Identity Politics, Immigration,"Mexican" Demonization and Hate-Crime. In *Proceedings of the International AAAI Conference on Web and Social Media* (Vol. 15, pp. 467-478). 10.1609/icwsm.v15i1.18076

PaquinV.FalardeauM. (2023). The complex impacts of climate change on ecosystems, food (in) security, and mental health. doi:10.31234/osf.io/dqjf6

Pareek, S., Bhatnagar, V., Poonia, R. C., Sharma, S., & Samanta, D. (2022). A Review on Rural Women's Entrepreneurship Using Machine Learning Models. *International Conference on Innovations in Computational Intelligence and Computer Vision*, 375–395.

Pariatam. (n.d.). Waster Management Developments in the last five decades. *The Journal of Sustainable Circular Economy.* doi:10.1177/0734242X231199938

Parmar, S., Kumar, M., & Kumar, A. (2022). The impact of Kerala floods on mental health: A cross-sectional study. *International Journal of Disaster Risk Reduction*, *70*, 102732.

Past climate changes and human adaptation. (2019, November). *Past Global Changes Magazine*, *27*(2). doi:10.22498/pages.27.2.83

Pauly, R., & SpringerLink (Online service). (2021). *Economic instability and stabilization policy: On the path from crises to state directed economies.* Springer Fachmedien Wiesbaden. doi:10.1007/978-3-658-33626-4

Pawaskar, P., & Penkar, S. a. (2018). Student's Perception on Social Media and Impact on their Education. *International Journal of Creative Research Thought*, *6*(1), 341–344.

Pee, L. G., & Pan, S. L. (2022). Climate-Intelligent Cities and Resilient Urbanisation: Challenges and Opportunities for Information Research. *International Journal of Information Management*, *63*, 102446. doi:10.1016/j.ijinfomgt.2021.102446

Peng, W. (2019). Construction and application of accounting computerization skills teaching resource database under the background of. *Curriculum and Teaching Methodology*, *2*(1), 1–4. doi:10.23977/curtm.2019.21001

Pepe, P. (2022, March). On Lyapunov Methods for Nonlinear Discrete-Time Switching Systems With Dwell-Time Ranges. *IEEE Transactions on Automatic Control*, *67*(3), 1574–1581. doi:10.1109/TAC.2021.3069661

Pihkala, P. (2020a, September 23). Anxiety and the Ecological Crisis: An Analysis of Eco-Anxiety and Climate Anxiety. *Sustainability (Basel), 12*(19), 7836. doi:10.3390/su12197836

Pihkala, P. (2020b, December 4). Eco-Anxiety and Environmental Education. *Sustainability (Basel), 12*(23), 10149. doi:10.3390/su122310149

PittMichaelAbbas. (2021)Public Interest Technology, Citizen Assemblies, and Performative Governance. *IEEE Technology and Society Magazine, 40*(3).

Podder, S. K., Thomas, B., & Samanta, D. (2022). Impact of Innovative Technology on Quality Education: The Resourceful Intelligence for Smart Society Development. *Machine Intelligence and Data Science Applications: Proceedings of MIDAS 2021, 132*, 293.

Podder, S. K. (2021). 2022). Impact of Innovative Technology on Quality Education: The Resourceful Intelligence for Smart Society Development. Machine Intelligence and Data Science Applications. *Proceedings of MIDAS, 132*, 293.

Poursadeqiyan. (2020). Migration Health Crisis Associated with Climate Change: A Systematic Review. *Journal of Education and Health Promotion, 9*(1), 97. doi:10.4103/jehp.jehp_4_20

Pradhan Mantri Krishi Sinchayee Yojana (PMKSY). (n.d.). *About PMKSY.* https://pmksy.gov.in/

Prerna, B. (2012). P., & Gupta, O.P. Multiple Sequence Alignment of Different Species. *Indian Journal of Applied Research, 1*(7), 78–82. doi:10.15373/2249555X/APR2012/24

Prieto Curiel, R., Cresci, S., Muntean, C. I., & Bishop, S. R. (2020). Crime and its fear in social media. *Palgrave Communications, 6*(1), 1–12. doi:10.1057/s41599-020-0430-7

Rabatseta, B., Maluleka, J. R., & Onyancha, O. B. (2021). Adoption and use of social media in academic libraries in South Africa. *South African Journal of Library and Information Science, 87*(1). Advance online publication. doi:10.7553/87-1-1926

RahamanM. A. (2023). *Climate Induced Migration and Population Dynamics of Bangladesh.* doi:10.13140/RG.2.2.13575.98729

Rai, A. (2023, August). Understanding Eco-Anxiety and Eco-Grief: A Comprehensive Review. *International Journal of Research Publication and Reviews, 4*(8), 681–686. doi:10.55248/gengpi.4.823.50353

Rajamani, S. K., & Iyer, R. S. (2023b). Networks in healthcare: A systematic review. *BioMedInformatics, 3*(2), 391–404. doi:10.3390/biomedinformatics3020026

Ramadan & Ataallah. (2021). Are Climate Change and Mental Health Correlated? *General Psychiatry, 34*(6). . doi:10.1136/gpsych-2021-100648

Rani, K., Vishnoi, H., & Mishra, M. (2023). A Mental Health Chatbot Delivering Cognitive Behavior Therapy and Remote Health Monitoring Using NLP And AI. *2023 International Conference on Disruptive Technologies (ICDT),* 313-317. 10.1109/ICDT57929.2023.10150665

Rasoolimanesh, S. M., Ramakrishna, S., Hall, C. M., Esfandiar, K., & Seyfi, S. (2020). A. *Journal of Sustainable Tourism*, 1–21. doi:10.1080/09669582.2020.1775621

Raulamo-Jurvanen, P., Mäntylä, M., & Garousi, V. (2017). Choosing the right test automation tool: a grey literature review of practitioner sources. *Proceedings of the 21st International Conference on Evaluation and Assessment in Software Engineering*. 10.1145/3084226.3084252

Ravinder, D. (2020). *Social Medias, Skills to Manage, Purposes, Advantages,and their Role in Library Sectors in Digital Era*. IJIRT.

Raz, S. (2022, July 4). Ethics of Climate Responsibility and Eco-Anxiety. *Psychoanalytic Dialogues*, *32*(4), 348–350. doi:10.1080/10481885.2022.2090814

Reed, P. (2022). Impact of social media use on executive function. *Computers in Human Behavior*.

Rehling, J. T. (2022, October 17). Conceptualising eco-anxiety using an existential framework. *South African Journal of Psychology. Suid-Afrikaanse Tydskrif vir Sielkunde*, *52*(4), 472–485. doi:10.1177/00812463221130898

Reser, J. P., & Swim, J. K. (2011). Adapting to and Coping with the Threat and Impacts of Climate Change. *The American Psychologist*, *66*(4), 277–289. doi:10.1037/a0023412 PMID:21553953

Rimal, Y., Rathore, V. S., Pageni, S., Samanta, D., Karuppiah, M., & Singh, D. (2022). Neural Network based Student Grade Prediction Model. *2022 International Conference on Machine Learning, Computer Systems and Security (MLCSS)*, 22–27. 10.1109/MLCSS57186.2022.00013

Ristea, A., Al Boni, M., Resch, B., Gerber, M. S., & Leitner, M. (2020). Spatial crime distribution and prediction for sporting events using social media. *International Journal of Geographical Information Science*, *34*(9), 1708–1739. doi:10.1080/13658816.2020.1719495 PMID:32939153

Ross, S., & Wall, G. (1999). Ecotourism: Towards congruence between theory and Practice. *Tourism Management*, *20*(1), 123–132. doi:10.1016/S0261-5177(98)00098-3

Rowe, F. (2018). Being critical is good, but better with philosophy: From digital transformation and values to future of IS research. *European Journal of Information Systems*, *27*(3), 380–393. doi:10.1080/0960085X.2018.1471789

Ruiz-Pérez, M., Maoyi, F., Xiaosheng, Y., & Belcher, B. (2001). Bamboo forestry in China: Toward environmentally friendly expansion. *Journal of Forestry*, *99*(7).

Rummukainen, M. (2012, February 16). Changes in climate and weather extremes in the 21st century. *Wiley Interdisciplinary Reviews: Climate Change*, *3*(2), 115–129. doi:10.1002/wcc.160

Saaid, S. A., & Wahab, Z. A. (2014). The impact of digital -based materials on Undergraduates' reading habit. *International Journal of Social Science and Humanity*, *4*(3), 249–253. doi:10.7763/IJSSH.2014.V4.357

Sahu, S. K., Chandra, J., Muloor, K., & Samanta, D. (2023). A Thorough Investigation of Various Goals and Responses for Mobile Software-Defined Networks. In *Designing and Developing Innovative Mobile Applications* (pp. 307–328). IGI Global. 10.4018/978-1-6684-8582-8.ch017

Salik, K. M. (2023). *A Threat or an Opportunity? Internal Migration in the Context of Climate Extremes in Pakistan.* doi:10.1007/978-3-031-34194-6_11

Samuel, M., & Shirley, C. P. (2023). Mindset, An Android-Based Mental Wellbeing Support Mobile Application. *2023 3rd International Conference on Pervasive Computing and Social Networking (ICPCSN),* 989-996. 10.1109/ICPCSN58827.2023.00169

Santhosh Kumar Rajamani & Radha Srinivasan Iyer. (2023a). Declining Freshwater Species Biodiversity. In N. K. Chourasia & K. Chahal (Eds.), *Perspectives on Global Biodiversity Scenarios and Environmental Services in the 21st Century* (pp. 22–56). IGI Global. doi:10.4018/978-1-6684-9034-1.ch002

Santhosh Kumar Rajamani & Radha Srinivasan Iyer. (2023c). Use of Python Modules in Ecological Research. In P. O. de Pablos, M. N. Almunawar, & M. Anshari (Eds.), *Perspectives on the Transition Toward Green and Climate Neutral Economies in Asia* (pp. 182–206). IGI Global. doi:10.4018/978-1-6684-8613-9.ch011

Sarma, E. A. S. (2004). Is Rural Economy Breaking Down? Farmers' Suicides in Andhra Pradesh. *Economic and Political Weekly, 39*(28), 3087–3089. https://www.jstor.org/stable/4415247

Saur Energy International. (2023). https://tinyurl.com/5b2hj7mz

Scheyvens, R. (1999). Ecotourism and the empowerment of local communities. *Tourism Management, 20*(2), 245–249. Advance online publication. doi:10.1016/S0261-5177(98)00069-7

Schwerdtle. (2018). The Health Impacts of Climate-Related Migration. *BMC Medicine, 16*(1). . doi:10.1186/s12916-017-0981-7

Scott, D., Peeters, P., & Gössling, S. (2010). Can tourism deliver its "aspirational" greenhouse gas emission reduction targets? *Journal of Sustainable Tourism, 18*(3), 393–408. doi:10.1080/09669581003653542

Scurlock, J. M. O., Dayton, D. C., & Hames, B. (2000). Bamboo: An overlooked biomass resource? *Biomass and Bioenergy, 19*(4), 229–244. doi:10.1016/S0961-9534(00)00038-6

Selby, J., Dahi, O. S., Fröhlich, C., & Hulme, M. (2017). Climate change and the Syrian civil war revisited. *Political Geography, 60*, 232–244. doi:10.1016/j.polgeo.2017.05.007

Self-Brown, S., Lai, B., Patterson, A., & Glasheen, T. (2017). The impact of natural disasters on youth: A focus on emerging research beyond internalizing disorders. *Current Psychiatry Reports, 19*(8), 1–7. doi:10.1007/s11920-017-0798-2 PMID:28681356

Sellers, S., Ebi, K. L., & Hess, J. (2019). Climate change, human health, and social stability: Addressing interlinkages. *Environmental Health Perspectives, 127*(04), 045002. doi:10.1289/EHP4534 PMID:30986089

Seraphin, H., Bah, M., Fyall, A., & Gowreesunkar, V. G. B. (2020). Tourism education in France and sustainable development goal 4 (quality education). *Worldwide Hospitality and Tourism Themes, 13*(1), 139–147. Advance online publication. doi:10.1108/WHATT-08-2020-0083

Shafer. (n.d.). *Climate Change Primer, Warm Heart World Wide.* https://tinyurl.com/mvpus57u

Sharma, D., Singh, S., Singh, B., Singh, A. V. J., Singh, R., Singh, R., Singh, G., Singh, J., Singh, B., Singh, R., & ... (2021). Farmer suicides in India: A policy-induced disaster of epic proportions. *Economic and Political Weekly, 56*(2), 13–16.

Sharma, M., & Angmo, R. (2014). Web based automation testing and tools. *International Journal of Computer Science and Information Technologies, 5*(1), 908–912.

Shaukat, K. (2015). Taxonomy of automated software testing tools. *International Journal of Computer Science and Innovation, 1*, 7–18.

Shaw, D. J. (2007). World food summit, 1996. In *World Food Security: A History since 1945* (pp. 347–360). Palgrave Macmillan UK., doi:10.1057/9780230589780_35

Sheffield & Landrigan. (2011). Global Climate Change and Children's Health: Threats and Strategies for Prevention. *Environmental Health Perspectives, 119*(3), 291–298. doi:10.1289/ehp.1002233

Shenoy, S. S., & Eeratta, R. (2011, December). Green software development model: An approach towards sustainable software development. In *2011 Annual IEEE India Conference* (pp. 1-6). IEEE. 10.1109/INDCON.2011.6139638

Shultz. (2018). Public Health and Mental Health Implications of Environmentally Induced Forced Migration. *Disaster Medicine and Public Health Preparedness, 13*(2), 116–122. doi:10.1017/dmp.2018.27

Silchenko, D., & Murray, U. (2023). Migration and climate change – The role of social protection. *Climate Risk Management, 39*, 100472. doi:10.1016/j.crm.2022.100472

Singh, P. K. (2020). Use of social media platforms with reference to library and information services in digital era. *BIMS International Research of Management and Commerce.*

Singh, S., Kakkar, P., & Bhambri, P. (2013). A Study of the Impact of Random Waypoint and Vector Mobility Models on Various Routing Protocols in MANET. *International Journal of Advances in Computing and Information Technology, 2*(3), 41–51.

Snyman, S. L. (2012). The role of tourism employment in Poverty Reduction and community perceptions of conservation and tourism in Southern Africa. *Journal of Sustainable Tourism, 20*(3), 395–416. doi:10.1080/09669582.2012.657202

Solomon, S., Plattner, G. K., Knutti, R., & Friedlingstein, P. (2009). Irreversible Climate Change Due to Carbon Dioxide Emissions. *Proceedings of the National Academy of Sciences of the United States of America, 106*(6), 1704–1709. doi:10.1073/pnas.0812721106 PMID:19179281

Soni, S. (2022). Impact of social media tools on enhancing library and information services.

Southwick, S. M., Bonanno, G. A., Masten, A. S., Panter-Brick, C., & Yehuda, R. (2014). Resilience definitions, theory, and challenges: Interdisciplinary perspectives. *European Journal of Psychotraumatology*, *5*(1), 25338. doi:10.3402/ejpt.v5.25338 PMID:25317257

Sri. (2023). Global Mental Health and Climate Change: A Geo-Psychiatry Perspective. *Asian Journal of Psychiatry, 84*, 103562. . doi:10.1016/j.ajp.2023.103562

Stubbs-Richardson, M., Rader, N. E., & Cosby, A. G. (2018). Tweeting rape culture: Examining portrayals of victim blaming in discussions of sexual assault cases on Twitter. *Feminism & Psychology*, *28*(1), 90–108. doi:10.1177/0959353517715874

Sundararaju, V. (2021, May 10). *A wonder bamboo from Tamil Nadu to mitigate climate change*. Down To Earth. https://www.downtoearth.org.in/blog/climate-change/a-wonder-bamboo-from-tamil-nadu-to-mitigate-climate-change-76851

Swachh Bharat Activities. (n.d.). swachhbharat.mygov.in

Syafiqah Ismail, I., Adilla Rashidi, N., & Yusup, S. (2022). Production and characterization of bamboo-based activated carbon through single-step H 3 PO 4 activation for CO 2 capture. *Environmental Science and Pollution Research International*, *29*, 1–7. doi:10.1007/s11356-021-15030-x/Published PMID:34189693

Talukder, B., Van Loon, G. W., Hipel, K. W., Chiotha, S., & Orbinski, J. (2021). Health impacts of climate change on smallholder farmers. *One Health*, *13*, 100258. doi:10.1016/j.onehlt.2021.100258 PMID:34027006

The Central Ground Water Board (CGWB). (n.d.). http://cgwb.gov.in/bhujal-news-0

The Ministry of New and Renewable Energy. (n.d.). *Annual Reports from 2016-2023*. https://mnre.gov.in/en/annual-report/

The Oxford Handbook of Acceptance and Commitment Therapy. (2023, June 20). Oxford University Press.

The WB IBRD. (n.d.). *IDA press release*. https://tinyurl.com/bp6y2au6

Tiwari, R. K., Kumar, A., & Joshi, P. K. (2021). Farmer suicides in India: A comprehensive analysis of official data (1995–2019). *World Development Perspectives*, *23*, 100345.

Torres, J. M., & Casey, J. A. (2017). The centrality of social ties to climate migration and mental health. *BMC Public Health*, *17*(1), 600. doi:10.1186/s12889-017-4508-0 PMID:28679398

Tricco, A. C., Lillie, E., Zarin, W., O'Brien, K. K., Colquhoun, H., Levac, D., Moher, D., Peters, M. D. J., Horsley, T., Weeks, L., Hempel, S., Akl, E. A., Chang, C., McGowan, J., Stewart, L., Hartling, L., Aldcroft, A., Wilson, M. G., Garritty, C., ... Straus, S. E. (2018). Straus, S. E.: Prisma extension for scoping reviews (PRISMA-SCR): Checklist and explanation. *Annals of Internal Medicine*, *169*(7), 467–473. doi:10.7326/M18-0850 PMID:30178033

Turner, C. (2018). *Global issue topics for essays and research.* Academic Press.

U.S. Committee for Refugees and Immigrants (USCRI). (n.d.). https://refugees.org/wp-content/uploads/2022/04/Mitigating-Mental-Health-Impacts-of-Climate-Related-Migration.pdf

U.S. Global Change Research Program. (2018). *Impacts, Risks, and Adaptation in the United States: Fourth National Climate Assessment* (Vol. II). Retrieved from https://nca2018.globalchange.gov/

Uddin. (2021). A Global Child Health Perspective on Climate Change, Migration and Human Rights. *Current Problems in Pediatric and Adolescent Health Care, 51*(6). . doi:10.1016/j.cppeds.2021.101029

UNHCR. (1984). *Cartagena Declaration on Refugees, adopted by the Colloquium on the International Protection of Refugees in Central America, Mexico and Panama, Cartagena de Indias, Columbia, 22 November 1984.* https://www.unhcr.org/media/cartagena-declaration-refugees-adopted-colloquium-international-protection-refugees-central

United Nation Framework Convention on Climate Change. (2008). *Report of the Conference of the Parties serving as the meeting of the Parties to the Kyoto Protocol on its third session, held in Bali from 3 to 15 December 2007.* https://unfccc.int/resource/docs/2007/cmp3/eng/09.pdf

United Nations Framework Convention on Climate Change (UNFCCC). (2008). *The Cancun Agreements: Outcome of the work of the Ad Hoc Working Group on Long-term Cooperative Action under the Convention.* https://unfccc.int/resource/docs/2010/cop16/eng/07a01.pdf

United Nations Framework Convention on Climate Change (UNFCCC). (2015). *Report of the Conference of the Parties on its 21st session, held in Paris from 30 November to 13 December 2015.* https://unfccc.int/resource/docs/2015/cop21/eng/10.pdf

United States Congress, Climate Change Science Program Office. (2008). *Analyses of the Effects of Global Change on Human Health and Welfare and Human Systems: Final Report, Synthesis and Assessment Product 4.6: Report by the U.S. Climate Change Science Program and the Subcommittee on Global Change Research.* U.S. Climate Change Science Program.

Uyanık, G. K., & Güler, N. (2013). A study on multiple linear regression analysis. *Procedia: Social and Behavioral Sciences, 106*, 234–240. doi:10.1016/j.sbspro.2013.12.027

Vafeiadou. (2023). The Influence of Climate Change on Mental Health in Populations of the Western Pacific Region: An Umbrella Scoping Review. *Heliyon, 9*(11). . doi:10.1016/j.heliyon.2023.e21457

van der Linden, S. L., Leiserowitz, A. A., Feinberg, G. D., & Maibach, E. W. (2015). How to Communicate the Scientific Consensus on Climate Change: Plain Facts, Pie Charts or Metaphors? *Climatic Change, 133*(3), 429–443.

Vehbi, B. O., & Doratli, N. (2010a). Assessing the impact of tourism on the physical environment of a small coastal town: Girne, Northern Cyprus. *European Planning Studies, 18*(9), 1485–1505. doi:10.1080/09654313.2010.492587

Verplanken, B., Marks, E., & Dobromir, A. I. (2020). On the nature of eco-anxiety: How constructive or unconstructive is habitual worry about global warming? *Journal of Environmental Psychology*, *72*, 101528. doi:10.1016/j.jenvp.2020.101528

Virani, J., Daredi, N., Bhanushali, A., Shukla, M., & Shah, P. (2023). Mental Healthcare Analysis using Power BI & Machine Learning. *2023 4th International Conference on Signal Processing and Communication (ICSPC),* 73-76. 10.1109/ICSPC57692.2023.10125680

Vishnu, J., & ... A strategy for optimal loading pattern of a typical power system — A case study. *2014 International Conference on Advances in Green Energy (ICAGE),* 112-117. 10.1109/ICAGE.2014.7050152

Von Braun, J., Swaminathan, M. S., & Rosegrant, M. W. (2005). *Agriculture, food security, nutrition and the Millennium Development Goals. In Annual Report Essays 2004EssayAR.* Intl Food Policy Res Inst.

Vorosmarty, C. J., Green, P., Salisbury, J., & Lammers, R. B. (2000). Global water resources: Vulnerability from climate change and population growth. *Science, 289*(5477), 284–288.

Wake, B. (2022, September 29). Understanding eco-anxiety. *Nature Climate Change, 12*(10), 884–884. doi:10.1038/s41558-022-01497-0

Walsh, J. P., & O'Connor, C. (2019). Social media and policing: A review of recent research. *Sociology Compass, 13*(1), e12648. doi:10.1111/soc4.12648

Wang, R., Guo, Z., Cai, C., Zhang, J., Bian, F., Sun, S., & Wang, Q. (2021). Practices and roles of bamboo industry development for alleviating poverty in China. *Clean Technologies and Environmental Policy, 23*(6), 1687–1699. doi:10.1007/s10098-021-02074-3

Water conservation under MGNREGA. (n.d.). https://tinyurl.com/yx6tcw72

Watson, E. (2020). *Education: The Potential Impact of Social Media and Hashtag Ideology on the Classroom.* Research in Social Sciences and Technology.

Web, A., Yusuke, T., Srilata, K., & Pradeep, K. (2023, July 19). *Notes from Tuvalu: Leading the way in adapting to sea-level rise.* PreventionWeb.https://www.preventionweb.net/news/notes-tuvalu-leading-way-adapting-sea-level-rise

Weber, E. U. (2010). What Shapes Perceptions of Climate Change? *Wiley Interdisciplinary Reviews: Climate Change, 1*(3), 332–342. doi:10.1002/wcc.41

Williams, M. L., Burnap, P., Javed, A., Liu, H., & Ozalp, S. (2020). Hate in the machine: Anti-Black and anti-Muslim social media posts as predictors of offline racially and religiously aggravated crime. *British Journal of Criminology, 60*(1), 93–117. doi:10.1093/bjc/azz064

Williams, M. S., & Patel, V. H. (2022). The impact of covid-19 on the mental health and well-being of immigrant healthcare workers. *Advances in Psychology, Mental Health, and Behavioral Studies*, 34–57. doi:10.4018/978-1-7998-8813-0.ch003

Willox, A. C., Harper, S. L., Ford, J. D., Edge, V. L., Landman, K., Houle, K., & Wolfrey, C. (2013). Climate change and mental health: An exploratory case study from Rigolet, Nunatsiavut, Canada. *Climatic Change, 121*(2), 255–270. doi:10.1007/s10584-013-0875-4

Winschiers, H., & Paterson, B. (2004). Sustainable software Sustainable Software development. *Proceedings of the 2004 annual research conference of the South Development African institute of computer scientists and information technologists on IT research in developing countries.*

World Economic Forum. (2022). *Global risks report 2023.* Retrieved April 27, 2023, from https://www.weforum.org/publications/global-risks-report-2023/in-full/1-global-risks-2023-today-s-crisis

World Food Programme. (2020). *WFP chief warns of hunger pandemic as COVID-19 spreads* (Statement to UN Security Council). Retrieved April 27, 2023, from https://www.wfp.org/news/wfp-chief-warns-hunger-pandemic-covid-19-spreads-statement-un-security-council

World Health Organization. (2022, June 16). *World Mental Health Report: Transforming mental health for all.* World Health Organization. https://www.who.int/publications/i/item/9789240049338

Wright. (2023). Climate Change and the Adverse Impact on the Health and Well- Being of Women and Girls from the Women's Health Expert Panel of the American Academy of Nursing. *Nursing Outlook, 71*(2). . doi:10.1016/j.outlook.2023.101919

Wysocki, W. P., Clark, L. G., Attigala, L., Ruiz-Sanchez, E., & Duvall, M. R. (2015). Evolution of the bamboos (Bambusoideae; Poaceae): A full plastome phylogenomic analysis. *BMC Evolutionary Biology, 15*(1), 50. Advance online publication. doi:10.1186/s12862-015-0321-5 PMID:25887467

Xu, X., & Reed, M. (2018). Perceived pollution and inbound tourism for Shanghai: A panel VAR approach. *Current Issues in Tourism, 22*(5), 601–614. doi:10.1080/13683500.2018.1504898

Yang, D., Heaney, T., Tonon, A., Wang, L., & Cudré-Mauroux, P. (2018). CrimeTelescope: Crime hotspot prediction based on urban and social media data fusion. *World Wide Web (Bussum), 21*(5), 1323–1347. doi:10.1007/s11280-017-0515-4

Yang, S. J., Zhang, Y. J., Goldstein, G., Sun, M., Ma, R. Y., & Cao, K. F. (2015). Determinants of water circulation in a woody bamboo species: Afternoon use and night- time recharge of culm water storage. *Tree Physiology, 35*(9), 964–974. doi:10.1093/treephys/tpv071 PMID:26232783

Yang, Y., Li, W., Zhang, Q., Zhang, L., Cheung, T., & Xiang, Y.-T. (2020). Mental health services for older adults in China during the COVID-19 outbreak. *The Lancet. Psychiatry, 7*(4), e19. doi:10.1016/S2215-0366(20)30079-1 PMID:32085843

Yap, C. T. M., & Yahya, M. Y. (2016). Compressive properties of Hawaiian gold timber bamboo under different conditions. *Journal of Advanced Research in Applied Mechanics, 25*(1), 10–18.

Yates, O. E. T., Groot, S., Manuela, S., & Neef, A. (2023). "There's so much more to that sinking island!" - Restorying migration from Kiribati and Tuvalu to Aotearoa New Zealand. *Journal of Community Psychology, 51*(3), 924–944. doi:10.1002/jcop.22928 PMID:36004412

Yonetani, T., & Gordon, H. B. (2001, February 8). Abrupt changes as indicators of decadal climate variability. *Climate Dynamics*, *17*(4), 249–258. doi:10.1007/PL00007926

Yuliia Matvieieva. (2021). *9 The Relationship Between Inequality and Socio-Economic Drivers as a Tool for Sustainable Territorial Development, in Economic Inequality – Trends, Traps and Trade-offs*. River Publishers.

Zea Escamilla, E., Habert, G., & Wohlmuth, E. (2016). When CO2 counts: Sustainability assessment of industrialized bamboo as an alternative for social housing programs in the Philippines. *Building and Environment*, *103*, 44–53. doi:10.1016/j.buildenv.2016.04.003

Zeng, C., Lyu, B., Deng, S., Yu, Y., Li, N., Lin, W., Li, D., & Chen, Q. (2020). Benefits of a three-day bamboo forest therapy session on the physiological responses of university students. *International Journal of Environmental Research and Public Health*, *17*(9), 3238. doi:10.3390/ijerph17093238 PMID:32384727

Zhang, N., Ren, R., Zhang, Q., & Zhang, T. (2020). Air pollution and tourism development: An interplay. *Annals of Tourism Research*, *85*(March), 103032. Advance online publication. doi:10.1016/j.annals.2020.103032

Zhong, S., Yang, L., Toloo, S., Wang, Z., Tong, S., Sun, X., & Huang, C. (2018). The long-term physical and psychological health impacts of flooding: A systematic mapping. *The Science of the Total Environment*, *626*, 165–194. doi:10.1016/j.scitotenv.2018.01.041 PMID:29339262

Zhu, L., & Zhang, Y. (2023). Fuzzy Judgment-based Mental Health Assessment System for College Students. *2023 IEEE International Conference on Integrated Circuits and Communication Systems (ICICACS)*, 1-5. 10.1109/ICICACS57338.2023.10099884

Zim, M. K. I., & Kaur, U. (2023). Mental health detection and anticipation analysis using Deep Learning. *2023 International Conference on Computer Communication and Informatics (ICCCI)*, 1-6. 10.1109/ICCCI56745.2023.10128581

About the Contributors

S. M. Aamir Ali is presently Assistant Professor at Symbiosis Law School Pune, Symbiosis International (Deemed University), Pune. He is currently pursuing his PhD from West Bengal National University of Juridical Sciences (WBNUJS), Kolkata, and has completed his LL.M from National Law School of India University (NLSIU), Bengaluru, with a specialization in Human Rights Law. Mr Ali has been teaching across specializations in Constitutional Law, Human Rights and Criminal Law at UG and PG levels since 2019.

Pankaj Bhambri is affiliated with the Department of Information Technology at Guru Nanak Dev Engineering College in Ludhiana. Additionally, he fulfills the role of the Institute's Coordinator for the Skill Enhancement Cell and acts as the Convener for his Departmental Board of Studies. He possesses nearly two decades of teaching experience. Dr. Bhambri acquired a Master of Technology degree in Computer Science and Engineering and a Bachelor of Engineering degree in Information Technology with Honours from I.K.G. Punjab Technical University in Jalandhar, India, and Dr. B.R. Ambedkar University in Agra, India, respectively. Dr. Bhambri obtained a Doctorate in Computer Science and Engineering from I.K.G. Punjab Technical University, located in Jalandhar, India. Over an extended period, he fulfilled many responsibilities including those of an Assistant Registrar (Academics), Member (Academic Council/BoS/DAB/RAC), Hostel Warden, APIO, and NSS Coordinator within his institution. His research work has been published in esteemed worldwide and national journals, as well as conference proceedings. Dr. Bhambri has made significant contributions to the academic field through his role as both an editor and author of various textbooks. Additionally, he has demonstrated his innovative thinking by filing several patents. Dr. Bhambri has received numerous prestigious awards from esteemed organizations in recognition of his exceptional achievements in both social and academic/research domains. These accolades include the ISTE Best Teacher Award in 2022 and 2023, the I2OR National Award in 2020, the Green ThinkerZ Top 100 International Distinguished Educators award in 2020, the I2OR Outstanding Educator Award in 2019, the

SAA Distinguished Alumni Award in 2012, the CIPS Rashtriya Rattan Award in 2008, the LCHC Best Teacher Award in 2007, and several other commendations from various government and non-profit entities. He has provided guidance and oversight for numerous research projects and dissertations at the undergraduate, postgraduate, and Ph.D. levels. He successfully organized a diverse range of educational programmes, securing financial backing from esteemed institutions such as the All India Council for Technical Education (AICTE), the Technical Education Quality Improvement Programme (TEQIP), among others. Dr. Bhambri's areas of interest encompass machine learning, bioinformatics, wireless sensor networks, and network security. Dr. Bhambri possesses a wide array of professional responsibilities, encompassing the duties of an educator, editor, author, reviewer, expert speaker, motivator, and technical committee member for esteemed national and worldwide organizations.

Anil Kumar Biswal is currently pursing PhD in Computer Science and Engineering from ITER, Siksha 'O' Anusandhan (Deemed to be University). He has also an extensive record of teaching for the past 7 years and his research interests include Wireless Sensor Networks, Cloud Computing and IoT. He has worked on various research projects that solve some social related issues like traffic with street light monitoring process, drowsiness detection system, designing women safety App, etc.

Anuttama Ghose is an Assistant Professor of Law and Program Co-ordinator of BA LLB (Hons.) Program at the School of Law, Dr. Vishwanath Karad MIT World Peace University, Pune, India. She is also presently Heading the Centre for Innovation Incubation and Intellectual Property Laws (CIIIPL) at MIT WPU School of Law. She has completed her B.A LL.B (IPR Hons.) from the School of Law, KIIT University, Bhubaneshwar and LL.M with specialisation in the field of Intellectual Property Laws from Symbiosis Law School, Pune, India. She has completed her PhD in the field of Protection of Non-Conventional Trademarks from School of Law and Justice, Adamas University, Kolkata. Dr. Ghose has further completed an online course on Copyright Laws conducted by Harvard Law School and Berkman Klein Centre for Internet & Society at Harvard University and an Academy Distance Course on IPR by the World Intellectual Property Organization (WIPO). She has also completed a Diploma Course in Entrepreneurship Administration and Business Law from West Bengal National University of Juridical Science, Kolkata. She has previously worked as Vice-Principal and Assistant Professor of Law at Indian Institute of Legal Studies, Darjeeling, West Bengal. Apart from being an IPR enthusiast, she is a very keen researcher and has various publications in National and International Journals to her credit.

Radha Srinivasan Iyer, a dedicated physiotherapist, specializes in Neurosciences, bringing her expertise to the forefront of pediatric care. With a passion for making a difference, she focuses on enhancing the lives of disabled children through tailored rehabilitation programs. Radha's compassionate approach and commitment to fostering mobility and independence make her an invaluable advocate for the well-being of her young patients. Through her work, she continues to leave a lasting impact on the lives of those with unique healthcare needs.

Suja John is a Professor of Tourism Studies with the School of Business and Management, CHRIST (Deemed to be University). She has been associated with Christ University for over 20 years in various capacities. To her credit, she has received copyright ownership for her work titled 'Bamboo the "Green Gold": A Strategic Resilient to Control Climate Change Issues Catalyst to attain Sustainable Development Goals. She has published two books, 'Community-Based Decentralized Waste Management for Sustainable Tourism' and 'Tourists Behaviour in the Age of Climate Change.' Dr Suja is a subject expert for the Tourism E-Content Development Project -National Mission on Education through ICT, Ministry of Human Resource Development, Government of India. As a part of the project, she developed two core papers: Hygiene and Sanitation and Geography for Tourism. She has also published research articles in International and national referred journals and participated and presented papers in various conferences and seminars in India and abroad.

Arya K. is a 2nd year BSc. Psychology student, studying at PES University, Bangalore, India.

Nethravathi K., M.Com, MBA, PGDBA, K-Set, Ph.D., is having a Professional Experience of 14 years in Education and a 1-year industry. She has a Ph.D. from CMR University. She has worked in various capacities such as HOD of Management and PG Coordinator and attended many workshops and Faculty Development Programs, seminars, and orientation programs. She has worked in various capacities as NAAC Coordinator, BBA Coordinator, Guide, and Mentor for three years. She has presented more than 35+ research articles at various National and International conferences. Out of which few were awarded the best research paper. She has traveled to various states to participate in workshops, conferences, and paper presentations. She has published many research articles in ISSN and ISBN, UGC care listed, Scopus indexed, ABDC, and published 3 Indian Patents and also has authored the textbook Accountant in Business. She has organized many offline and online FDPs and SDPs and is well appreciated as a Resource Person on various topics of Research Methodologies and Finance and Taxation. She is currently working in the Department of Business Management Studies, and guiding research scholars in the School of Commerce, Jain Deemed to be University.

Zidan Kachhi is a counselling psychologist and an assistant professor. He has authored two books and published various papers in peer reviewed journals.

Jyoti Khandelwal, an Assistant Professor at the University of Engineering & Management in Jaipur, holds a Ph.D. in Information System Security from Manipal University Jaipur. With 12 years of teaching experience, she earned her B.Tech and M.Tech in Computer Science. Jyoti has published nine papers in SCI and Scopus indexed journals, showcasing her dedication to advancing knowledge in cybersecurity. Her passion for the subject and extensive teaching experience make her a valuable mentor for students, blending practical insights with academic excellence.

K. V. N. Lakshmi is an Assistant Prof. in Vijaya College, having a strong academic and teaching background, possessing a very good experience in degree programmes nearly 13years of significant experience demonstrating excellent teaching that fosters students critical skills leading to significantly improved performance, comfortable with in all teaching environment and able to demonstrate transferable skills.

Smitha Amarail Mohanan is a full-time research scholar at CHRIST (Deemed to be University), Bangalore, India. She specialized in Sustainable Tourism Development and Sustainable Development Glos area and is doing PhD in the area of Sustainable Tourism development and Women Empowerment. She has received Indian copyright ownership for her work titled 'Bamboo the "Green Gold": A Strategic Resilient to Control Climate Change Issues Catalyst to attain Sustainable Development Goals. She has been invited by Pacific Asia Travel Association (PATA) as a key speaker in the webinars-Focused Youth Studies on the topic "Bamboo the "Green Gold": A Strategic Resilient to Control Pollution and Community Development in the tourism sector." She has participated in four international conferences and presented papers. Smitha's vision is to help rural women who live in vulnerable situations through tourism by seeking the possibilities and opportunities in the field that would enable them to have a good quality of life.

Eswara Reddy Orekanti has completed his Ph.D. from Jawaharlal Nehru Technological University Hyderabad, INDIA. He has a rich experience of 24 years both in teaching and industry. Dr. Reddy served in different academic capacities in various institutions. Currently, serving at Mohan Babu University, Tirupati, Andhra Pradesh, INDIA in the capacity of Professor and BOS Chairman, Department of Civil Engineering. He has published more than 50 papers in reputed journals and conferences. He has 7 patents filed/published to his credit. He is a Life Fellow/Life Member of many professional bodies.

Digvijay Pandey is a lecturer for the Department of technical Education, Research Scholar IET Lucknow, India.

Revathi Pasupuleti has completed her M.Tech. from Jawaharlal Nehru Technological University, Anantapur, India. Presently, she is working as an Assistant Professor at PVKK Institute of Technology. Subsequently, she is pursuing her Part-time Ph.D. in Civil Engineering at Mohan Babu University, Tirupati, Andhra Pradesh, India.

Amrita Raj is an M.Tech scholar, and passionately delves into the intricacies of their field, merging theoretical knowledge with practical applications. Their academic journey is marked by a relentless pursuit of excellence, evidenced by their dedication to cutting-edge research. Fueled by curiosity, they are committed to contributing valuable insights to the technological landscape.

Santhosh Rajamani is an ENT professor at MIMER Medical College in Pune, India. He has over 15 years of teaching and clinical experience and has published extensively on ENT topics. Dr. Rajamani has won several gold medals and awards for his academic achievements. He is an editorial member for multiple medical journals and serves as a reviewer. Dr. Rajamani is passionate about research and advancing ENT care.

Kamalraj Ramalingam received his Bachelor of Engineering in CSE in the year of 2002 from Bharathiyar University, Coimbatore. In the year of 2009 he has completed his Master of Engineering in CSE from Anna University, Chennai. After then, he received PhD degree from Anna University in the year of 2017. He is having more than 19 years of experience in teaching field. During that he has presented many papers in National and International conferences and he published papers in International Journals and patents also. He is a member of in 'Indian Society of Technical Education'. Now currently he is working as a Professor in School of Computer Science & IT in Jain (Deemed to be) Univeristy, Bangalore, Karnataka, India.

Sita Rani is a faculty member of the Dept of Computer Science and Engineering at Guru Nanak Dev Engineering College, Ludhiana. She earned her Ph.D. in Computer Science and Engineering from I.K. Gujral Punjab Technical University, Kapurthala, Punjab in 2018 and has more than 20 years of teaching experience to her credit. She has completed Post Graduate Certificate Program in Data Science and Machine Learning from Indian Institute of Technology, Roorkee in 2023. She has also completed her Postdoc from Big Data Mining and Machine Learning Lab,

South Ural State University, Russia in August, 2023. She is an active member of ISTE, IEEE, and IAEngg and is the recipient of the ISTE Section Best Teacher Award 2020, and the International Young Scientist Award 2021. Dr. Rani has contributed to various research activities while publishing articles in renowned journals and conference proceedings. She has published three international patents and has delivered many expert talks in A.I.C.T.E. sponsored Faculty Development Programs along with organizing many International Conferences. Her research interest includes Parallel and Distributed Computing, Machine Learning, the Internet of Things (IoT), Healthcare, and Digital Twin.

Shantha V. is a 2nd year BSc. Psychology student, studying at PES University, Bangalore, India.

Index

P

R

S

T

U

V

W

Y

www.ingramcontent.com/pod-product-compliance
Lightning Source LLC
LaVergne TN
LVHW061540290926
849162LV00004B/862